Saving for Retirement

Saving for Retirement

Intention, Context, and Behavior

Gordon L. Clark, Kendra Strauss, and Janelle Knox-Hayes

UNIVERSITY PRESS

HUMBER LIBRARIES LAKESHORE CAMPUS
3199 Lakeshore Blvd West
TORONTO, ON.   M8V 1K8

OXFORD
UNIVERSITY PRESS

Great Clarendon Street, Oxford OX2 6DP

Oxford University Press is a department of the University of Oxford.
It furthers the University's objective of excellence in research, scholarship,
and education by publishing worldwide in

Oxford New York

Auckland Cape Town Dar es Salaam Hong Kong Karachi
Kuala Lumpur Madrid Melbourne Mexico City Nairobi
New Delhi Shanghai Taipei Toronto

With offices in

Argentina Austria Brazil Chile Czech Republic France Greece
Guatemala Hungary Italy Japan Poland Portugal Singapore
South Korea Switzerland Thailand Turkey Ukraine Vietnam

Oxford is a registered trade mark of Oxford University Press
in the UK and in certain other countries

Published in the United States
by Oxford University Press Inc., New York

© Gordon L. Clark, Kendra Strauss, and Janelle Knox-Hayes, 2012

The moral rights of the authors have been asserted
Database right Oxford University Press (maker)

First published 2012

All rights reserved. No part of this publication may be reproduced,
stored in a retrieval system, or transmitted, in any form or by any means,
without the prior permission in writing of Oxford University Press,
or as expressly permitted by law, or under terms agreed with the appropriate
reprographics rights organization. Enquiries concerning reproduction
outside the scope of the above should be sent to the Rights Department,
Oxford University Press, at the address above

You must not circulate this book in any other binding or cover
and you must impose the same condition on any acquirer

British Library Cataloguing in Publication Data
Data available

Library of Congress Cataloging in Publication Data
Data available

Typeset by SPI Publisher Services, Pondicherry, India
Printed in Great Britain on acid-free paper by
MPG Books Group, Bodmin and King's Lynn

ISBN 978–0–19–960085–4

1 3 5 7 9 10 8 6 4 2

For the memory of Bryan Victor Clark and James Christopher Knox...

Contents

Preface ix
Acknowledgments xv
List of Figures xviii
List of Tables xix

1. Introduction 1
2. Environment and Behavior 17
3. Risk Propensities 39
4. Sophistication, Salience, and Scale 59
5. Being in the Market 76
6. Housing, Retirement Saving, and Risk Aversion 97
7. The Demand for Annuities 118
8. The "New" Paternalism 136
9. Pension Adequacy and Sustainability 153

Bibliography 172
Author Index 191
Subject Index 195

Preface

Pension policy has changed dramatically over the past two decades. In the postwar period, pension policy was largely concerned with the distribution of risk between the state and the employer in the context of a multipillar system in which the first (state-sponsored) and second (occupational or employer-sponsored) pillars dominated. During the 1950s and 1960s, the zenith of the breadwinner welfare state in many advanced Western economies, pensions were part of a social compact that had its foundation in a tripartite set of seemingly stable institutions: the Keynesian state, the standard employment contract, and the nuclear family headed by a male breadwinner. While these institutions, and their related norms and conventions, were never a "complete" blueprint, they nevertheless set the parameters of many people's lives.

After the oil shocks and recession of the 1970s, the landscape started to change, particularly in Anglo-American countries: those characterized by Gosta Esping-Andersen among others as "liberal" (compared to "social democratic"). Keynesianism was swept aside by the reemergence of liberalism in its late twentieth-century guise. As Jamie Peck among others have noted, neoliberalism is associated with global finance, the discounting of state power in the context of globalization, a premium on social and economic individualization, and labor market flexibility. The forces of neoliberalism have profoundly affected the three key institutions of the postwar welfare state underpinning retirement income. Global finance has challenged governments' inherited commitments to retirement income, global trade and exchange have undercut the welfare state, the renewed focus on individual rights and entitlements has been accompanied by greater exposure to market risks, and labor market flexibility has seen the demise of the standard contract of employment and the single employer career.

As societies have changed, so have the pressures on pension systems increased. For occupational pensions, this has produced a marked shift in the allocation of risk, choice, and responsibility from the firm to the individual, especially in the United States and the United Kingdom. It is clear that the traditional defined benefit (final salary) pension is in terminal decline in the private sector in these countries, and under pressure even in the social democratic and conservative-corporate states such as Sweden, Norway, and

Preface

Germany. This type of pension scheme has been replaced in the United States and elsewhere by the defined contribution (DC) plan. In DC plans the final pension is determined by a combination of contributions and investment returns, minus charges; there is no guaranteed level of benefits and the risks are almost entirely borne by the individual plan member, although in some jurisdictions (e.g., Sweden) there is a guaranteed minimum rate of return in state-sponsored plans. Occupational pension policy, in countries like the United Kingdom and the United States, has shifted away from issues of collective risk sharing and management toward individual capacity and responsibility in areas such as financial decision-making and literacy.

The rationale for this book is grounded in the apparent transformation of the means by which people save for retirement. While we have doubts about the DC model, and the ability of DC occupational pensions to provide an adequate retirement income for many of today's workers, we recognize that DC pensions are a significant and growing feature of the pensions landscape. To not engage with the principles, logics, and the model of human behavior that underlies them is to cede the debate about the future of pensions to those who downplay their risks. Moreover, as the subdiscipline of behavioral economics gains attention and influence with its own multifaceted critique of the strong model of economic rationality that dominates orthodoxy, the importance of engaging critically with the field becomes ever more evident. For while a reinvigorated behavioralism has been hugely important in pointing out the limitations of *homo economicus*, its own assumptions about the nature of human decision-making denuded of social identity and meaning deserves greater scrutiny.

The behavioral revolution has opened a door for social scientists hitherto closed for at least a couple of generations. This revolution has undercut the hegemony of the rational actor model which is deeply embedded in social science theory and practice, particularly but not solely in the field of economics. Just ten years ago it would have been difficult if not impossible to publish papers and books that challenged the status quo, let alone suggest alternative conceptions of intention, behavior, and social relationships. The behavioral revolution is associated with Herbert Simon, Daniel Kahneman and Amos Tversky, Gerd Gigerenzer, and many others besides. Their research has been inspiring, not least because of their persistence. The behavioral revolution is anything but a sudden rupture with the past and the imposition of a new order. The revolution has been coming for many years as evidenced by the papers on the topic published by Herbert Simon in the 1950s but otherwise ignored until he was awarded the Nobel Prize in economics.

The significance of the behavioral revolution has been underscored by the global financial crisis. While there have been other crises, prompted by events such as the failure of Long Term Capital Management in the late 1990s, the

Preface

global financial crisis has challenged core beliefs in the intellectual foundations of financial economics. Most importantly, doubts have been raised about the plausibility of the efficient markets hypothesis, the presumed discipline associated with self-interest, and the assumption that financial innovation always adds to social well-being. We have witnessed an enormous destruction of wealth and pension savings around the world, with implications for the retirement welfare of many millions of people. Understanding the nature and scope of, and limits to, financial decision-making is crucial if the lessons of the global financial crisis are to be realized in more effective institutions and a more secure retirement for future generations.

In this context, the behavioral revolution provides welcome instruments for interrogating conventional expectations as regards individual financial decision-making. Our book is about planning (intention) and decision-making (behavior) with specific reference to saving for retirement. We are motivated by a concern about the many people whose eventual retirement is reliant upon their accumulated savings inside and outside DC pension plans. We are not only interested in saving for retirement; we are also interested in how people conceptualize the planning process given their social backgrounds and relationships, where they live, and the cultural milieu in which they make (or do not make) plans for the future. It is our contention that the context of intention can have an important role in affecting people's ambitions in saving for retirement. The interaction between context, social identity, and behavior is an aspect of decision-making that remains undertheorized in the new behavioralism.

The behavioral revolution is, of course, very much concerned with individual decision-making in market economies. Financial markets are characterized by significant levels of risk and uncertainty and have proved to be anything other than stable and predictable over the past few decades. In many respects, behavioral theorists take for granted this type of environment, focusing on observed patterns of behavior and especially systematic deviations from the expectations associated with the rational actor model. Elsewhere we have utilized this research, investigating the decision-making performance of those charged with the responsibility of making savings and investment decisions on others' behalf. Here we suggest that theorists and policymakers should take more seriously the behavioral responses to these types of environments—an argument that is hardly contentious given the global financial crisis. How people respond to market imperatives and the social and geographical patterns of income inequalities produced by those responses are important for future retirement welfare.

We are less concerned with generating new findings about the nature of cognition than we are concerned with better understanding behavior in context. This is, of course, a long-running program research in the social

Preface

sciences and especially economic geography. In this case, we take seriously Herbert Simon's scissors metaphor: when describing the interaction between cognition and context (what he also referred to as the environment in which people behave), he suggested that a pair of scissors represents the type of interaction he believed produces behavior. That is, one blade of the scissors represents our cognitive capacities and predispositions, while the other blade of the scissors represents the environment in which we plan and act, including the formal and informal institutional norms and customs that may reward or indeed sanction certain types of behavior. While Simon is quite properly identified as one of the founding fathers of the behavioral revolution, it should also be remembered that he was a remarkable and insightful analyst of institutional structure.

Our book builds upon the insights and observations of those who led the behavioral revolution in the social sciences. At the same time, it works away from those insights to incorporate the environment into our understanding of savings behavior. This is a broader conception of intention, behavior, and relationships that invokes the embedded nature of cultural norms as well as where people live and work. This is an ambitious research program, one that is not so easily charted in terms of accepted protocols and modes of research. As we work away from the behavioral revolution to a more social and geographical perspective on savings intentions, we do so by making associations and demonstrating relationships rather than testing, in the manner of experimental psychology, for fundamental human traits. Consequently, our book can be seen as one step toward a research program that seeks to map Simon's scissors by reference to the scope of human behavior. As such, we are very much concerned with the scale at which behavior takes place and the scale at which cues from the environment are incorporated into intentions.

Early in the book we review the relevant behavioral research on myopia. In doing so, we recognize that there is an enormous body of research focused upon *temporal* myopia: the predisposition of human beings in favor of immediate events and the presumption that human beings heavily discount the future. Myopia has, of course, another dimension: the apparent predisposition of human beings in favor of the *local* rather than the global, and the difficulty many of us have in assessing the significance of events in certain times and places *against* social and economic processes that operate over extended times and at higher spatial scales. People are not necessarily trapped in space–time myopia. Whether by habit, intuition, or the imitation of others, people look for ways to manage these predispositions if not consciously then at least in terms of the associated and often implicit rewards (and sanctions) that go with certain types of behavior. We are at pains throughout the book to show that behavioral responses to risk and uncertainty are socially stratified and that myopia can be regulated by the relationships and social resources that people have to hand.

In sum, our book is doubly unorthodox. Recognizing and relying upon the insights generated by the behavioral revolution in relation to the limits of the rational actor model, we seek to show that understanding how people save for retirement must be sensitive to the other neglected blade of Simon's scissors: the environment in which people live and work. This includes the relationships so important to their emotional lives, and the institutional rewards and sanctions that give life or otherwise to a sense of consciousness about what might be achieved by deliberate thought and planning.

In a nutshell, our destination is the context of decision-making (the "environment" of choice) rather than cognition. As a consequence, the research tools we use do not map on the experimental procedures of cognitive psychology. To bring context into the story about saving for retirement we have had to combine standard empirical strategies with a commitment to measure something that seems almost impossible to measure: the interaction between social identity, cultural milieu, and where people live with expectations about future retirement welfare and their relationships with others. One of the advantages of beginning from economic geography is that we are able to capitalize on its ambivalence about rationality and its unease about universal models of behavior dislocated from the time and place wherein relationships are formed and expectations generated. The challenge, of course, is to make sense of the complex interplay between behavior and the environment in ways that provide points of leverage or significance for the design of social institutions.

Acknowledgments

We were initiated into the world of financial decision-making some years ago through research sponsored by the UK National Association of Pension Funds on the competence and consistency of pension fund trustee decision-making. In this research program we were able to develop research protocols and tests of decision-making with the insights and encouragement of the late John C. Marshall. Not only did this research program underwrite Emiko Caerlewy-Smith's DPhil, it helped a great deal in our application to the UK Economic and Social Research Council (ESRC) for a DPhil CASE award to support Kendra Strauss's research into the financial decision-making of defined contribution pension plan participants.

The ESRC award was made possible by our partnership with Mercer Human Resources Consulting, and especially the support of Divyesh Hindocha. Throughout the project he was encouraging and helpful, applying the resources of Mercer to underwrite the research program. This included access to Mercer's national survey on pension benefits and our engagement with an unnamed investment bank located in London which allowed us to survey their defined contribution plan participants on issues related to housing, saving for retirement, and pensions. We would also like to acknowledge the contributions made by Roger Urwin, Jonathan Gardner, and Michael Orszag of Towers Watson for their interest in our research and, in Roger's case, his role as co-author in Chapter 9 of the book. Towers Watson also provided a database on defined contribution pension plan participants. This was especially useful in prompting Janelle Knox-Hayes' interest in the area. There is no doubt that industry support of our research made the book possible. In Janelle's case, we would also like to acknowledge the Jack Kent Cooke Foundation for support of her MSc and DPhil at Oxford.

The research reported in this book has been presented in a number of venues, including annual conferences of the Association of American Geographers and the Royal Geographical Society, the Vega Symposium (Stockholm) honoring Allen Scott, seminars at Oxford University, Birkbeck College, Warwick University, Westminster University, the Oxford and Southampton Institutes of Ageing, the Allianz-Oxford Pensions Conferences, and industry events sponsored by Mercer and Towers Watson. We would like to thank

Acknowledgments

these organizations for the opportunity to present our research and hear back from the audience comments and criticisms that have added to the final product.

Of the many people who have helped the research process we would like to acknowledge the inspiration of the late John C. Marshall—he is greatly missed. Research assistance was provided, in part, by Roberto Duran-Fernandez and Oliver Cover from Oxford's Department of Politics as well as Stephen Almond; their expert advice and computation skills added a great deal to this book. Both Roberto and Stephen were also co-authors on a couple of papers that form the basis of this book. Insights and comments on the various papers that are the basis of the chapters of this book were provided by Jeremy Cooper, James Churcher, Adam Dixon, Don Ezra, Paul Gerrans, Howell Jackson, David Knox, Ashby Monk, Courtney Monk, Linda McDowell, John Livanas, Olivia Mitchell, Alicia Munnell, Douglas Hershey, Ron Martin, Peter Sunley, Susan Smith, Maria Evandrou, Athina Vlachantoni, Jane Falkingham, Ewald Engelen, Paul Langley, Amelia Sharman, and Neil Wrigley. Crucially, Olga Thönissen, Brenda Stones, and Jan Burke provided the assistance necessary to realize the book project. Finishing touches to the manuscript were done while the first-named author was a guest of Professor Allan Fels, Dean, Australian and New Zealand School of Government, Melbourne.

This book is based, in part, on a series of papers published over the past few years on pension saving and behavior in a variety of venues. The authors would like to acknowledge Elsevier for permission to publish portions of "Financial knowledge" from the *International Encyclopaedia of Human Geography* edited by Rob Kitchin and Nigel Thrift (2009) [978-0080449111] with John C. Marshall; John Wiley for permission to publish portions of "Human nature, the environment, and behavior: explaining the scope and geographical scale of financial decision-making," *Geografiska Annaler B* (2010) 92: 159–73 [ISSN 0435-3684] and "Mapping UK pension benefits and the intended purchase of annuities in the aftermath of the 1990s stock market bubble," *Transactions of the Institute of British Geographers* (2007) NS32: 539–55 [ISSN 0020-2754]; Cambridge University Press for permission to publish portions of "Individual pension-related risk propensities: the effects of socio-demographic characteristics and a spousal pension entitlement on risk attitudes," *Ageing and Society* (2008) 28: 847–74 [ISSN 0144-686X]; Pion Ltd for permission to publish portions of "Financial sophistication, salience, and the scale of deliberation in UK retirement planning," *Environment and Planning A* (2009) 41: 2496–515 [ISSN 0308-518X]; Oxford University Press for permission to publish portions of "'Being in the market': the UK house-price bubble, and the intended structure of individual pension investment portfolios," *Journal of Economic Geography* (2010) 10: 331–59 [ISSN 1468-2702] with Roberto Durán-Fernández; Palgrave Macmillan for permission to publish portions of "The 'new'

Acknowledgments

paternalism, consultation and consent: expectations of UK participants in defined contribution and self-directed retirement savings schemes," *Pensions: An International Journal* (2009) 14(1): 58–74 [ISSN 1478-5315]; Sage for permission to publish portions of "The home, pension savings and risk aversion: intentions of the defined contribution pension plan participants of a London-based investment bank at the peak of the bubble," *Urban Studies* (2011) DOI: 10. 1177/0042098011410332 [ISSN 0042-0980] with Stephen Almond; finally we would like to thank the International Society of Certified Employee Benefit Specialists for permission to reproduce portions of our paper "DC pension fund best-practice design and governance" (with Roger Urwin) published in the 4th Quarter 2011 issue of *Benefits Quarterly* [ISSN: 8756-1263] devoted to global benefits.

As is customary, none of the above should be held responsible for any views or opinions expressed herein. This is especially true for our industry sponsors and supporters.

Oxford, August 2011

List of Figures

3.1 Asset-allocation choices for selected socio-demographic groups   55

4.1 Classification of planning types with reference to the salience of issues and the sophistication of judgment   63

5.1 Behavioral predispositions of market agents at the intersection between time and space with respect to financial decision-making under risk and uncertainty   80

8.1 Blackstone's target date fund   140

List of Tables

3.1	Socio-demographic characteristics of the analysis sample	47
3.2	Marginal probabilities on risk aversion in the probit model	51
4.1	Probabilities of significant variables in the ordinal logistic regression of planning importance	69
4.2	Statistically significant correlates of pension planning; parameter signs, strength of significance, and consistency across the indicators of planning capacity	71
5.1	Correlation matrices of macro-variables, quarter-to-quarter: 1988–2007	86
5.2	Property as a retirement savings instrument (company pension scheme)	91
6.1	Results of ordinal logistic regression for attitudes to property as a form of retirement saving	109
6.2	Results of ordinal logistic regression for attitudes to property as a form of retirement saving, including attitude to financial risk as a predictor	112
8.1	Level of consultation desired by income	146

1

Introduction

Occupational pensions are a relatively recent innovation. Their history is well documented: see, for example, Ginn et al. (2001), Munnell and Sundén (2004), and Thane (2006). Our intention is not to provide another account of their genesis. Rather, we begin the book with the issue of *why* occupational pension systems came into being, what their current purpose is, and whether they meet the (relatively modest) goals as set out in many governments' policy manifestos. This forms the backdrop for our critical examination of financial decision-making, and in particular saving for retirement. The implications for pension policy and institutions are examined in the concluding chapter of the book.

It is important to distinguish between the rationale for a public state pension, and the rationale for an occupational pension system. Although the two are related, they are nevertheless distinct in their genesis and *raison d'être* as well as in their subsequent evolution. The concept of a state-funded old age pension was first mooted to deal with the problem of the impoverished elderly in the industrial societies of the late nineteenth century. As Thane (2006) points out, prior to the mid-twentieth century the assumption in most countries was that families and communities would support themselves through work and thrift. In this sense, the collapse of traditional networks and institutions of support related to processes of industrialization and waged-work that created new forms of poverty in old age—and the proposed solutions to deal with them—were intimately bound up with notions of morality and concerns about incentives to save and work.

According to Sass (2006: 79), the British civil service pension plan of 1859 "became the model in the use of old-age pensions for developing a career managerial workforce." What is important about this observation is that the goal of this, and other similar models of pension provision, was workforce management and control. Especially in the Anglo-American world, occupational pensions were conceived as an enterprise benefit rather than a welfare benefit, and a tool for managing staff retention and development (Standing, 2009). This model of supplementary income, linked to the evolution of white and blue collar work, was fundamentally related to the development of a standard model of employment in industrial societies. This model of employment, which reached its zenith in the post-WWII period, was associated with Fordist and later Taylorist methods of production, large vertically integrated firms, Keynesian economic and social policy, and a male-breadwinner model of economy and society.

The standard contract of employment, which underpinned this model, was never universal. Yet management practices and the structure of the archetypical industrial firm provided incentives for corporations and companies to bond workers to their firms through pay and benefit structures, promotion pathways (generally internal labor markets), pension plans, and promises of job security (Stone, 2005). Meanwhile for labor, the era was typified by negative freedoms and partial or, according to Standing (2009), fictitious decommodification facilitated in part by the state: freedom from job insecurity, freedom from destitution in retirement. These negative freedoms were part of the social compact, which favored relative stability in labor relations, strong national-level institutions and production regimes, and a stable share of national income for labor (so as to encourage consumption).

The unraveling of this social compact has been associated, in relation to labor and employment, with the rise of neoliberal government policy and forms of governance, a new era of globalization driven by technology and finance, the global integration of low-wage economies from the global "south" (especially China, India, and Brazil), the construction of flexible labor markets, and welfare state retrenchment. It has also been typified by increased social and cultural tensions around migration, changing family and community structures, and changing demography (ageing societies). Individual agency and responsibility has been articulated as a set of freedoms, positive and negative, *against* the constraints of the traditional welfare state: freedom to change jobs and occupations, freedom to choose how one's assets are invested, freedom from taxation, freedom to choose where and how traditional services (education, health, social care) are delivered (see Beck and Beck-Gernsheim, 2001; Charles and Harris, 2007; Frericks et al., 2007).

A significant feature of this change in orientation has been an increasing emphasis on the importance of occupational pensions for an "adequate" retirement (although it is worth remembering that the UK government always understood the potential role of occupational pensions in deflecting pressure for improving the rate of the minimalist state). Blair's Labour government, for example, made a commitment to shift from an average of 40 percent (now) to 60 percent of income from private pensions by 2050, reversing the historic dependence on the Basic State Pension (see DSS, 1998; Foster, 2010; Waine, 2009). And yet the affirming of this policy commitment has coincided with the decline in the scope and adequacy of occupational pensions for most employees in the private sector. If this target is to be met, it will have to come from increasing the coverage rate and value of defined contribution pensions in the private sector.

Neoliberalism, Rationality, and Choice

Liberalism has thus reemerged as a core political ethos prescribing the respective responsibilities of the individual, the community, and the state in relation to saving for retirement. The liberalism associated with John Stuart Mill combined a commitment to individual autonomy and responsibility with a recognition that not all people were equipped by reason of intelligence, education, and moral character to be fully responsible. By contrast, neoliberalism has had little patience with state paternalism; individuals are deemed fully responsible for their well-being in the global economy (Peck, 2010).

With neoliberalism came the hegemony of a certain type of economic theory predicated upon individual rationality and expressed through utility maximization. Being both a recipe for theory and a normative vision for society, this type of economics has been associated with the deregulation of financial markets, the fostering of shareholder (as opposed to stakeholder) value, and the global integration of trade and financial transactions. When considered over the broad sweep of the twentieth century, and especially developments after the First World War, individuals are increasingly responsible for their own long-term welfare—in part, a function of their performance in financial markets. And yet, at its moment of victory, the hegemony of the rational actor model has slipped away in the face of the critique from behavioral finance and the global financial crisis. Cognitive science, with its empirical focus and exacting testing protocols, has raised significant doubts about the capacity of people to play the role assigned them by neoliberalism (Langley, 2008). The global financial crisis has given great credence to this research agenda.[1]

It is important to remember that post-Enlightenment visions of rationality have long been critiqued from a variety of epistemological, ontological, and political positions: the Cartesian duality between the (rational) mind and (irrational, emotional) body, the association of the rational mind with the male gender and the body/emotions with women, and the translation of

---

[1] In fact, it is arguable (as many have suggested) that the global financial crisis is a test of the rational actor model and its doppelgänger the efficient markets hypothesis *and* is an expression of the hegemony of the efficient markets hypothesis for market theorists and policymakers alike. See the 2008 comments before Congress of Alan Greenspan on the assumptions made by the US Federal Reserve Bank regarding the unlikely prospect that rational market agents would act in ways that would be self-defeating. Events have proven otherwise. See Lee et al. (2009) for more details.

rationality into calculation and utility maximization at the hands of economic theorists. Feminists have been at the vanguard of related critiques, just as radical political economists have challenged the plausibility of the whole edifice. But these forms of critique made little impression on orthodoxy, whether focused on the economics discipline or on its more general model of human behavior. Economic geography has also sought to engage with and criticize a priori presumptions in favor of utility maximization and asocial, aspatial conceptions of autonomous agents (see, e.g., Barnes, 1988; McDowell, 2005; Peet, 2000; Strauss, 2008*b*). These engagements have, however, tended to be one-sided.

If people are to be the rational actors of conventional theory and policy, however, they need to plan and act in ways consistent with their own interests and in relation to the interests of others in a well-functioning market economy. Given their dependence upon the burgeoning financial markets of the late twentieth and early twenty-first centuries, individuals have required a certain quality and quantity of financial information and knowledge, such that their decision-making is consistent with their long-term (if not short-term) well-being. These ideas rest on a number of assumptions: individual utility maximization is always consistent with the interests of the actor; all agents have equal access to timely, accurate, and comprehensible information; and markets always function efficiently.

By this logic, financial decision-making is the bedrock of neoliberalism; it is a necessary ingredient in any government policy that attempts to shift responsibility for social welfare to the individual. As such, financial decision-making is a matter of substance (e.g., knowing how markets function), and a matter of social identity and even morality (being responsible, informed citizens willing to play the role assigned to them by society at large). But, of course, in reality most people are social beings with access to, at best, imperfect information, and markets never preexist human behavior and cooperation in some "pure," efficient form. Markets are the products of aggregate behavior, political, social, and economic norms and institutions, and the geometries of power.

In what follows, we sketch our approach to understanding the nature and scope of behavior giving due recognition to the themes of the book: cognition and context. In doing so, we follow the lead of Bernard Williams (1995: ch. 7) and others to the effect that understanding the nature and scope of human behavior is an empirical, as well as a political–philosophical, project. We eschew models of behavior that assume people are simply calculating machines, processing information from the environment against prior commitments through cognitive dispositions. While we take seriously our biological heritage, we recognize that the realization of intentions through behavior is not preordained by ironclad biological imperatives that ride roughshod over the circumstances that affect and even structure our lives.

Introduction

In this chapter, we identify the larger threads of argument in the behavioral literature, thereby providing a way of placing our own approach to understanding the determinants of saving for retirement in relation to the extant theories. Given the significance we attribute to social attributes and relationships in subsequent chapters, it should not be surprising that we hope to move beyond the stripped-down versions of the rational actor model that dominates in mainstream social science and often shapes the design of pension institutions and policy, in favor of a perspective that takes seriously the context in which people find themselves. The relevance of these themes will become more obvious in subsequent chapters.

Scope and Significance

Financial decision-making clearly matters in many aspects of everyday life, some of which are of a short-term nature while others have profound long-term consequences for welfare. Financial decision-making can be deployed for rudimentary decisions, such as discriminating between consumption items on the basis of cost and quality. So, for example, when we assess washing machines and dryers, the conventional model of behavior suggests that we should do so by comparing the attributes of various options assuming a budget constraint and some base-line preferences of expected use and location in the home (but see Iyengar, 2010 on choice overload). As summarized, this assessment process could be quite formal and explicit. In fact, the evidence suggests that people use shortcuts or heuristics, as well as social cues, to sort the available options into manageable packets where, in the last instance, issues of cost and value may be used as tie-breakers between equally desirable products. To think otherwise would be to deny the power of advertising! In these situations, financial knowledge may not be necessary to make a decision; all that may be required is a tried-and-tested decision rule and a well-honed understanding of social and cultural codes.

There are, of course, other more important decisions that cannot be easily reversed once taken, where costs cannot be recouped in any simple sense. For example, in many countries people purchase healthcare insurance where the long-term costs of underinsurance may be so significant that short-term cost-effectiveness is ignored. Likewise, in many countries, the middle class are significant consumers of private education for their children; here, cost-effectiveness must be balanced against emotional commitment, cultural capital, and (perhaps) the expected long-term rate of return of human capital. In these situations, where past decisions affect long-term welfare, financial decision-making is intimately related to individuals' discount functions—the value attributed to short-term consumption against expected long-term well-being. People might reasonably apply some rudimentary financial decision technique in all the above cases. But buying a washer-dryer is different in *type* from buying healthcare, education, and pensions.

The distinction drawn here is between financial decision-making that is relatively shallow, being contingent on current information about short-term benefits, and financial decision-making that is relatively deep, being contingent upon long-term expected prospects and their consequences. The former refers to decisions that are typically reversible (albeit at some cost), whereas the latter refers to knowledge and decisions that are in large part irreversible (albeit with exceptions in some cases).[2] It can be observed that the former is almost always about events within individuals' background circumstances, where the only knowledge needed to make a decision is the information inherited from past decisions (e.g., that other products of a particular manufacturer have been reliable and good value for money). The latter may require knowledge only partially available from past decisions and is, more often than not, dependent upon financial expertise not shared widely through society. For example, calibrating the risk-adjusted rate of return on one kind of healthcare policy over other kinds of healthcare policies requires a wide range of detailed knowledge (including knowledge about the relationship between healthcare and life-time earnings).

Whether about rudimentary or complex time-dependent issues, financial information must be assembled and then applied through informal or formal decision processes and routines. In a sense, financial decision-making is more valued for its instrumental use than for its intrinsic quality. This suggests two implications. First, individuals must decide on how much information and of what quality to collect, recognizing that the costs of assembling data must be balanced against its value in decision-making. This may be an instinctive rather than a reasoned decision—one of the reasons that governments, banks, and other institutions provide information on gathering and using information (through financial literacy and capability programs). Second, given the costs of decision-making (time, effort, and other opportunities foregone), people have incentives to apply existing decision templates rather than approach each decision afresh on its own merits. Inevitably, much of financial decision-making is based on intuition and habit in case-specific circumstances (Hogarth, 2001). There are significant incentives for the vendors of financial products to blur these distinctions in the hope of cultivating immediate consumption.

---

[2] Recent research on the costs of deep, long-lasting decisions, including those that involve retirement saving, suggests that where we begin from in time and space can make a profound difference to both the path of asset accumulation *and* the end-result. In these circumstances, initial decisions on issues such as asset allocation and savings vehicles can make a profound difference to the end result—a fact of life that cannot be reversed (we cannot rerun our lives) although we may live to regret those decisions and even the cohort to which we belong.

Introduction

Rationality and Knowledge

Rationality, defined as a basic set of cognitive capabilities, is a universal characteristic of humans whatever their context or culture. But individuals who share a basic cognitive capacity as human beings also vary considerably in terms of their cognitive performance, including their ability to assess and evaluate alternatives according to the dictates for formal and/or economic logic. The latter, moreover, is likely to be the product of socialization and education rather than innate ability. More significantly what counts as sensible (or acceptable) in some societies will not necessarily be the same in other societies; context and culture (being here synonyms for environmental factors) are crucial when people evaluate or judge behavior against social standards.[3] Recognizing that human beings share cognitive capabilities may be an *anti-essentialist* position, *contra* arguments about the inherent superiority of a racial or ethnic group or gender, and/or about geographical determinism. It is not the same as claiming that all people all the time behave in the same way in the same situation or that a particular socioeconomic system (e.g., capitalism) is "natural."

Rationality represented by the subjective expected utility (SEU) maximization model makes two specific assumptions about the utilization of financial decision rules. First, given the price of knowledge, it is rational to economize on its collection and use; second, outcomes (positive and negative) are symmetrical in that they are equally valued for their consequences.

Unfortunately for the SEU model, neither of these assumptions holds true in real life. It is self-evident that agents vary in terms of the financial decision-making they can afford. The knowledge they can afford affects the options they consider, and the options considered may be suboptimal in terms of the maximization of individual and/or collective (household, community) welfare. With experience calibrated on past decision metrics, individuals may become isolated from the best options and even from the better options they can reasonably afford. This type of behavior may be legitimated by cultural preferences which serve to justify favoring some options over other options. It is also apparent that many people, whatever their socio-demographic status, are risk averse, preferring the certainty of a known but small "win" over a much larger but risky potential "win" (Kahneman and Tversky, 1979). Being risk averse may be more or less valued in different settings, attracting the admiration of some and the approbation of others. These cultural cues are likely to constrain, reinforce, or, in some instances, determine behavior (see Chapter 2 for exposition of this point).

---

[3] There is increasing research on the interplay between cognition, culture, and context, especially as regards the importance of social standards in judging risk. See the remarkable cross-nation experimental study led by Henrich et al. (2005).

Most importantly, SEU models of rationality have been criticized for their shallowness regarding social identity and the significance of the emotions. It is frequently assumed that social identity and the emotions adversely affect reason because they filter what is observed and the implications to be drawn there from. Therefore, to the extent that reason is informed by information and financial decision-making, social commitments and emotions are thought to prompt wishful thinking and biased or capricious decision-making. However, cognitive psychologists have sought to counter these assumptions with empirical evidence suggesting that the emotions may be a valuable intuitive device for first-order responses to changing circumstances. This supports the positions of those in the social sciences who contend that it is impossible to separate out the domains of emotion and calculation. If financial decision-making has a formal quality such that it is a means of assessing the virtues or otherwise of competing options, the emotions may also provide a simple mechanism for presorting options by salience. By this account, explicit financial decision-making could be added to the mix of decision techniques at the end of a sequence of more intuitive judgments, rather than being located at the start of the process.

In practice, people are more or less rational, in the "strong" sense. They approach problems from intersecting vantage points: their intrinsic cognitive capacity *and* their relationships, sociocultural ties, and experience in certain settings or environments (as suggested by Herbert Simon's 1956 metaphor of scissors). People are also innately emotional in the sense that they bring to situations intuitive judgments based upon fear, anger, happiness, and love (see *Emotional Knowledge* section). For some theorists, however, financial decision-making is properly *the* antidote to the confounding affects of emotion and commitment. But this seems rather utopian (at best) and normative (at worst): it is hard to imagine a situation where individual behavior would not have cultural and social significance including, crucially, emotional significance, making the model quite possibly irrelevant. Being *purely* rational is surely a normative statement of supposedly logical behavior stripped bare of emotional commitment rather than a statement of lived-life.[4] It is also often an ideologically driven claim about proper (even laudable) behavior in the context of market-based societies.

---

[4] It is surely inconceivable that people's behavior can be distinguished by various components, as if the emotional can be separated from the analytical, etc. No doubt people may seek to distinguish between their emotional commitments and their economic interests (for example). But this is surely an act of conscious will not behavior that relies upon the actual or hypothetical separation of these "functions" in the brain. See Ortony et al. (1988) on the cognitive emotions.

Introduction

Risk, Uncertainty, and Scale

Financial markets are awash with information. There are information markets for vendors and for buyers. In fact, the problem is not so much the lack of information as the lack of means by which to discern relevant information and our inability to discriminate between information and information sources in terms of their relevance to market conditions. Cost and power asymmetries in access are also important discriminating factors. Directly or indirectly, financial products involve risk and uncertainty.

So, in the context of recent events in local and global financial markets, the purchase of a home mortgage may involve the vendor and buyer in complex time-dependent calculations of expected inflation and interest rates. Vendors who underestimate interest rates over a specified period of time effectively subsidize mortgagees (and vice versa). Moreover, many people do not appropriately weight low-incidence but costly risks. Continuing the mortgage example, during the 1980s and 1990s many UK home-buyers took out endowment mortgages, betting that a low-risk event (collapse of the London stock market) would not occur or would not be compounded by collateral threats to household income and wealth. More recently, vendors were willing to assume risks that were neither adequately priced nor understood in the mistaken belief that some other institutions, somewhere else, had made those assessments and could bear the downside risks.

Many people are unaware of the risks they face in everyday life. They certainly do not appreciate the fact that risks are related, such that one event may cascade to affect a person's entire well-being. To deal with these issues, governments have vigorously encouraged disclosure policies, transparency in product design and management, and the adoption of risk-assessment by formal decision trees that deliberately expose contingent risks (see Chapter 2, *Reconceptualizing Personhood* section). Most importantly, governments have encouraged "plain English" disclosure policies, using terms and concepts anchored by average rather than expert competence. But as suggested above, these policies may be compromised on three counts. First, rationality is idealized, eschewing the complications of context and culture. Second, rationality is portrayed as a logical process rather than an issue of substance. Third, rationality is conceived as unbounded, notwithstanding the fact that people more often than not are "myopic"; that is, they limit the scope of problems rather than see them whole.

"Naïve" investors are led to assume that the available information is representative of stable economic processes. Financial information is taken at face value with little appreciation of the motives of those who produce and market such information. In part, this reflects the lack of alternatives, because independent, cost-effective expert information is costly to obtain. By necessity, risk is calibrated using past information and extrapolated into the future. Therefore, naïve investors tend to use financial information in ways that reinforce past commitments; information at odds with individual predispositions is often discarded or ignored, suggesting that the costs of assimilating such information are such that it is easier to wait for exogenous events to force through reconciliation or a change in tactics. In practice, naïve investors tend to associate cost with reputation; rightly or wrongly, the presumption is that the higher the cost of financial information the more reliable the information. This point is elaborated in Chapters 4 and 5.

"Sophisticated" investors ideally scrutinize the integrity of financial information, seeking evidence of contamination by competing interests.[5] Rather than extrapolating from surface trends, they look for changes in the variance of underlying time-series, believing that this is one indication of market instability. Sophisticated investors believe that uncertainty is endemic; finely calibrated risk profiles of financial products, the core of advertising programs by the finance industry, suggest a level of certainty that is not justified by the performance of financial markets. For sophisticated investors, financial judgment is more important than financial information. But there is a paradox: financial decision-making must be continuously assessed in the light of changing circumstances. It must be constructed and deconstructed at every turn.

There is nevertheless a geographical scale-effect implied by this distinction between "naïve" and "sophisticated" investors. The former draw on that which is immediately to hand, extrapolating from what they know to the immediate future, often ignorant of the causal links between current events, future prospects, and their relationships here and there (local and global). As we explain in subsequent chapters, "myopic" behavior is inevitably, though not exclusively, local in the sense that naïve planners are not Bayesian analysts.[6] By contrast, it

---

[5] The distinction between "naïve" and "sophisticated" investors is made by a number of authors, including Stein (2009). In his case, it is suggested that individual investors are naïve, whereas institutional investors are sophisticated; the distinction rests on the apparent advantages of the latter in quantitative modeling based upon "academic research in finance" and their use of leverage in making the most of market movements. Stein suggests that sophisticated investors are "rational arbitrageurs," whereas individuals are not. We are not convinced that categorical distinctions such as rational versus irrational make a great deal of sense; it is probably better to talk about competence and expertise than rationality per se. In any event, considering the failure of many institutions to manage their own positions in the lead-up to the global financial crisis, it is difficult to accept that institutions are *ipso facto* rational in the sense that they are better able to conceptualize and realize coherent investment strategies.

[6] This discussion implies that being myopic is costly and, in a sense, self-defeating. We should acknowledge that focusing upon the local as opposed to the global, and the immediate future as opposed to the long-term, may be an effective strategy to control that which can be controlled (and action taken in regard thereof). Williams (1995: 208) is sympathetic to these coping strategies, suggesting that local knowledge is often more complex and multivalent than acknowledged by theorists.

Introduction

would seem that sophisticated planners are able to look into the future with a sense of the causal linkages that join observed events with underlying processes. By implication, they are able to operate locally and globally, as in fact they may have to, given the integration of financial markets around the world. In this respect, they can be deemed Bayesian by impulse and application.

Stylized Facts

It has become apparent that people are not the efficient information-processing machines of economic theory or, for that matter, government policy (although we should recognize that government policy is hardly ever coherent on this issue given the vestiges of paternalism that infuse the modern nation-state). On average, people are risk averse, eschewing financial opportunities that a so-called fully rational person would or should assume. On average, people are inefficient users of information, often backing ill-informed opinion when they should collect more information, or perversely collecting more information than is warranted by the scope of the problem. On average, people overvalue the near-future and undervalue the long-term future. On average, people do not carry through on past plans, "jumping at shadows" when they ought to stand by informed commitments. On average, people are poor at data analysis, befuddled by even the most elementary notions of probability and contingent risk (as illustrated in Clark et al., 2006, 2007).

But the results of cognitive science do not tell the whole story. Research on the financial behavior and planning of different sorts of people suggests a more nuanced picture. While the precise details vary by country, especially if we are to include developed economies outside the OECD, we can crudely suggest that financial literacy is distributed on a 20/60/20 basis.[7] That is, 20 percent of the population appears to have such poor financial competence, knowledge, and access to advice that their immediate and long-term welfare is imperiled. At the other end of the spectrum, 20 percent of the population can be characterized as sophisticated investors, combining access to expertise with knowledge and understanding of the nature and performance of financial markets. In between are perhaps 60 percent of the population, who are, at best, naïve investors subject to many of the cognitive shortcomings noted above, but with the expectations and resources to be active consumers of financial products.

---

[7] The 20/60/20 formula is provided so as to prompt consideration about the scope of competence in the population, and to represent our intuition about the facts of the matter in Western economies. We may be quite wrong, but which way? Is the bottom 20 percent really 30 percent? And does the distribution vary over time, such that in episodes of market volatility the top 20 percent shrivels to 5 percent?

It is tempting, therefore, to correlate cognitive ability in relation to financial decision-making with age, gender, and socioeconomic status. Sociodemographic status counts because status is a good proxy for income and educational resources (what Pierre Bourdieu calls *cultural capital*)—the assets needed to acquire financial knowledge and expertise so as to compensate for acknowledged cognitive shortcomings. Studies have also shown that risk aversion, for example, is frequently correlated with income and household wealth, which chimes with the concentration of equity ownership along class lines (notwithstanding the "shareholder society" supposedly heralded by the dot.com boom) and differential capacities for hedging against shocks and uncertainty. The *interaction* of socio-demographic characteristics with financial decision-making is understudied, perhaps because the testing regimes of cognitive psychologists often involve groups of undergraduates and MBA students from elite universities. This is one of the motivating forces behind our book: we show, in fact, that socio-demographic status counts in making plans for the future and that status can be an important means of discriminating between naïve and sophisticated decision-makers.

Most importantly, it appears that formal education, professional qualifications, and task-specific training can make a difference to people's financial expertise. University-level education in subjects demanding quantitative skills and the attainment of professional qualifications post-university in areas of related knowledge do make a difference, and long-term task-specific training seems to reinforce the advantages of education and professional qualifications. Inevitably, education, professional qualifications, and training are correlated, albeit imperfectly, with household income. It is also apparent that, notwithstanding the massive postwar improvement in educational attainment, in Western societies, high-quality education, quantitative skills, and training are socially stratified and not widely distributed. This has implications for inequality in societies where social and economic welfare is increasingly subject to the performance of financial markets.

Context of Knowing

By this account, effective financial planning is intimately related to conceptual and analytical sophistication. Just as plainly, the quality and quantity of financial information available to individuals is related to their socioeconomic status. There also appears to be a relationship between the quality of financial planning and expertise, education, qualifications, and training. Most importantly, many people have neither the cognitive ability nor the acquired skills to be expert decision-makers. They must therefore rely upon others through social relationships and networks to improve the acquisition and use of financial decision-making. The evidence suggests, in fact, that most people depend upon family when seeking to extend their financial decision-making and better calibrate their decision-making. By this logic, the household is a very important resource for information, advice, and decision cues. It need not contain any more expertise than the sum of its parts; nonetheless, it may be able to rule out extremes and fanciful assumptions.

But small groups may also be subject to the influence of dominant individuals (whatever their expertise or lack thereof) and may converge on conventional or even catastrophic solutions rather than the best or even second-best solutions. This is an argument for broadening the planning environment to include the workplace and wider sections of society at large. Some thirty or forty years ago, the workplace combined paternalism with class-specific entitlements, such that the available advice and knowledge were allocated by virtue of job classification. However, evidence on the value of workplace information exchange is mixed; surveys suggest that employees rarely share financial information with one another. More often than not, employers provide the relevant information and brief programs on the issues and options. While employees often indicate that this information is useful, sign-up rates for information briefings are reported to be low, with default options dominating deliberate employee choice. Research on pension plan participants has shown that the behavioral effects of such training programs are limited, despite positive attitudinal responses.

Another source of financial planning is peer-imitation and advertising. If conventional class-related social aspirations are less meaningful in (post)modern societies, it is apparent that many people make financial decisions according to their role models (public figures) and their peers (represented in the media by television programs and the like). In part, planning-by-imitation may be driven by workplace relationships and professional ties. In part, planning-by-imitation may signal aspirations of social identity and recognition. Consequently, financial decision-making may be nothing special; the consumption (the accumulation and assessment) of information for financial decision-making is likely to be governed by the same behavioral customs and conventions that govern the purchase of fashion items such as clothes and cars, where people live, and the organizations to which people belong. It is hard to believe otherwise.

In this world of image, media, and consumption, the Internet and Web-based resources are sometimes identified as important alternative (good and bad) sources of financial information (Clark et al., 2004). It appears that access to this medium is heavily dependent upon an individual's work environment and job tasks. It also appears that younger people are more willing to use the Web to search for information and to make purchasing decisions based on web information than older people (a finding that will not surprise those knowledgeable of the adoption of innovation). However, most websites, especially those sponsored by government and not-for-profit consumer advocates, utilize an abstract display format rather than a media-driven format, emphasizing a claimed difference between (important) financial decision-making and other kinds of (less important) consumer information. It is notable that our research suggests that people do not value "remote" sources of information; they prefer information and insight that can be gleaned through "local" relationships, whether by trust or contract (see Chapter 4).

Emotional Knowledge

The "contextual" model of financial planning suggests that it is made at the intersection between individual cognition and the environment (the family, place of work, society at large, etc.). By this logic, the exercise of cognition provides incomplete solutions if not supplemented by specific learning (formal and informal). But there is another sense in which cognition is incomplete: the emotions also play a significant role in determining behavior. To agree with this statement is to move a significant step away from the rational actor paradigm that dominated the social sciences over the second half of the twentieth century and embrace a twenty-first-century theory of the mind-brain that stresses cognitive integration and functional interdependence. The interaction of individual cognition and the environment also suggests that the rational individual is not an island: cultural and societal interactions, personal relationships, and environmental and geographical factors matter. Thus, it provides a clue as to the nature of intuition—the moment when people reach a solution by some "unconscious" mix of cognitive response, social resources, and emotion.

Although emotional expression is modulated by social learning, basic emotions have a built-in physiological origin. Emotions such as fear, disgust, happiness, and anger are mediated by innate brain circuits in higher primates, humans included. Other emotions like love are even more modulated by culture and learning and more ascriptive in origin; ascribed behavior draws social approbation or admiration (including imitation). Following Aristotle, we can argue that the emotions may be thought to be a form of visceral judgment made about a situation and the behavior that follows. In this sense, the emotions "value" information by visceral judgment, providing a response that may or may not be overruled by deliberation and the application of social resources. This way of integrating the emotions with decision-making and behavior recognizes that most behavior is cognitively responsive rather than consciously reflective.[8]

---

[8] Folk-law in the finance industry would have it that men and women are different in that the former are analytical and the latter are emotional (left brain/right brain dominant, etc.). This claim

So, for example, fear may be the response of many people to the financial market crisis. As a basic emotion, but with a large overlay of acquired triggers and meanings, it represents foreboding about a sudden reversal of fortune bringing to the fore the likely negative consequences of such an event. It may follow a stock-market bubble characterized by euphoria and the suspension of disbelief ("irrational exuberance"), and be prompted by deep-seated emotions from our biological past, but applied to very different material circumstances. It is little wonder, therefore, that risk aversion is so systematic among humans across the whole range of experience, whether they are educated or not, skilled or not, or reflective or not. Taking this one step further, fear and happiness may (respectively) amplify or discount adverse financial information, just as these emotions may select among the available financial information to fashion an interpretation of current circumstances. Fear and happiness are shared human experiences that lead to "herd behavior" (analogous to Akerlof and Shiller's 2009 "animal spirits").

For those less convinced about the role and significance of the emotions in financial planning, consider a more social emotion: love. Clearly, love is associated with biological drives such as sexual desire and reproduction. On the other hand, love is also associated with socially desired behavior such as care for the young, companionship, respect for the old, and a commitment to the welfare of family and friends. Not all people are bound by love; some are entirely selfish (self-love). However, the comparison is revealing. Those committed to others may have very shallow discount functions, whereas those who are wholly selfish may have very steep discount functions. Love may give rise to hyperbolic discount functions, whereas selfishness may be manifest as weakness of will. As feminists have long pointed out, caring behavior is essential for both social reproduction and social cohesion. Either way, people's need or otherwise for approval may be a powerful determinant of how they respond to financial markets.

Looking Forward

Financial planning is a field of research and an area of public policy that is bound to grow in significance over the coming years, notwithstanding the global financial crisis. The consequent retrenchment of inherited state institutions and commitments is likely to move Western societies further in the direction of dependence upon global financial markets (somewhat paradoxical given the public costs of the financial crisis). The

---

is empirically ill-founded and makes gross generalizations about people from specific environments with distinctive and hierarchical segmentation of tasks and functions.

place of financial planning at the intersection between the retreating state and the growing responsibilities of individuals for their own welfare suggests that it is one of the most important determinants of long-term social welfare. At the same time, so many of us appear to be naïve decision-makers and investors; we lack the skills of financial professionals and will never have enough information and knowledge to make up the difference. The relative inability of people to manage shortfalls in decision-making, the costs and benefits of social resources, and the play of emotions suggest that some market agents will benefit from exploiting these dichotomies. Here lies an issue of political economy, in that the welfare consequences of such inadequacies in the context of the global financial crisis are bound to flow back into the political arena.

As described, financial planning is vital to the well-being of adult men and women. Less understood are the possible pay-offs of financial education and financial literacy programs aimed at young children and teenagers. In play, in this regard, are a series of complex issues having to do with the cognitive development of children as they mature to adulthood and the degree to which they are able to assimilate information from the environment in their putative decision frameworks. Also at stake is the political project of creating "investor subjects," ideologically acclimatized to the individualization of risk (Langley, 2006). At the other end of the age distribution there are issues of similar significance, notably the consequences of ageing for cognitive functioning and the capacity of the elderly to respond to changing circumstances. At its core, liberalism values human autonomy and responsibility whatever people's ages. But there are counterarguments for the desirability of some level of paternalism, especially on the side of those who are simply not able to play the roles assigned to them by neoliberalism.

In any event, there remains a most troubling issue: how individual choice through financial decision-making is exercised in the context of culture and emotion. It was suggested above that intuition is one expression of their interrelationship. It was also suggested that focus on cognition at the expense of cultural expectations and emotional commitments would be incomplete as a matter of empirical reality and as a matter of relevance. While easily expressed as such, there are few frameworks available that allow for a means of integration. Furthermore, given the significance of financial information and knowledge for well-being, the academic literature has preferred silence on these issues to a fully fledged research program that would set the shape of things to come.

2

Environment and Behavior

Explaining behavior often invokes dualisms such as environment and behavior, society and space, and economy and culture. In each, there tends to be an implied tension between structure and agency: individual behavior is deemed dependent or contingent on the external world. The antimonies at work in such pairings suggest, as well, juxtaposed or competing modes of human life. That they are joined together indicates a lack of confidence in the explanatory power of just one element of a pairing (such as economy compared to culture) but recognition, nonetheless, that their pairing, in some sense, adds value. Presumably, a close reading of their interaction can provide us with the means of understanding the diverse landscape of human activity. So, for example, "relational economic geography" seeks to combine market imperatives with the fabric of social relationships to provide a deeper understanding of geographical differentiation and path dependence (see Bathelt, 2006; Bathelt and Glückler, 2005).

A number of disciplines are preoccupied by questions of emphasis and balance, straddling the divide between formal models of market structure and performance and broader conceptions of human life. Granovetter's economic sociology (1985) is an attempt at reconciliation with applications to conventional topics like job search behavior. In anthropology, culture has played a vital role in explaining differentiation between the peoples of the world being, at times, invoked as the essential building block of any society (for a critique, see Kuper, 2000). Economic geography has been more circumspect than anthropology and more like economic sociology, holding to universal models of human behavior seen through the lens of fine-grained studies about the significance of the context in which people live and work (Peck, 2005). But this strategy can be accused of obfuscation (at best) or cowardice (at worst): it does not tackle the integrity of the rational actor model that underpins most models of behavior.

Here, we sketch a means of integrating environment and behavior without capitulating to naïve economism or grandiose gesture. Our book is informed by research on finance and pension welfare at the interface between social science and public policy. This program is summarized in Ameriks and Mitchell (2008) and concerned with financial decision-making under risk and uncertainty where respondents have immediate needs and concerns as well as a long-term interest in their material well-being. In part, research is based on psychological experiments and test regimes designed to assess decision-making in theory and practice. As shown in this book, we can also use econometric methods to demonstrate the significance of the socio-demographic attributes of people in risk-related decision-making. In carrying through the research developed in this book, we have been inspired by the work of Kahneman and Tversky (1979). Even so, their project has its limits: they are firmly entrenched in cognition and test response rather than context, intention, and behavior.

It was noted in Chapter 1 that we rely upon Herbert Simon's scissors metaphor (1956) to the effect that human behavior occurs at the intersection between cognition and the context or environment in which people live and work.[1] To reap the full benefits of his insights on human behavior, we need a complete theory of cognition, a well-specified notion of context, and a theory of their interaction amendable to experimental analysis. Of course, such an integrated theory of behavior does not exist: it is at the frontier of research in psychology and the social sciences (Henrich et al., 2005).

This chapter works from first principles: human nature and the scope of rationality. We downplay deliberation and calibrated judgment in favor of customary modes of behavior: intuition, habit, and imitation. A way of conceptualizing the environment or context is suggested with reference to culture and society, although it is also argued that a truly *geographical* conception of context would be explicit about the *scale* of effect. To illustrate, we refer to recent published research on financial decision-making and summarize the findings and logic of our research program on pension planning for the scale and scope of decision-making. In conclusion, we ask whether the notion of "personhood" embedded in much of the literature is adequate for understanding the life-course of people who are the "objects" of research.

Rationality, behavior, and the role of environment in shaping human life are contentious issues that have preoccupied the social sciences for many decades. The apparent generality of these issues is betrayed by profound disagreement on even the simplest issues, such as the definition of cognition, context, and the environment. In the Appendix, brief comments are provided on relevant definitions. Notice, this chapter is not intended to resolve debate

---

[1] As such, our book seeks an expansive conception of behavior, sensitive to diversity and affect. By contrast, neuroeconomics is reductionist in intention, treating economics as a natural science whose object is to develop the theory of rational choice and thereby explain behavior (see Padoa-Schioppa, 2008).

once and for all over their meaning or for that matter their relationships to one another. It is no exaggeration to say that their definition is enmeshed in the much larger debate about human nature. This issue, though, has proved toxic for many disciplines, including geography. And yet, because of its claim of engagement with nature and society, it would seem incumbent upon human geography to have an explanation of behavior that is richer than the cardboard image of rational human nature.

By contrast, much of decision theory is focused upon sketching a parsimonious theory of behavior, or the standards by which to assess such a theory (see, e.g., Bermúdez, 2009). Rarely does decision theory seek to explain the observed diversity of behavior. Too often people are treated as one-dimensional fixed entities. If appropriate for theory-building, there are problems with such an approach, not least of which is the often unquestioned assumption that people have constant utility functions—especially over time, but also over space, in the sense that experience is discounted as relevant to the formation and reformation of a person's goals and objectives.

Human Nature

We could assume that people are basically selfish, acquisitive, and immoral; or we could assume that people are basically sociable, cooperative, and caring of others. The first "model" of human nature provides a rationale for designing institutions that protect us from one another; whereas the second "model" of human nature provides a rationale for institutions that are a framework for social action and the advancement of public welfare. Both models are universal in their scope, and are essentially normative in the attribution of intentions and behavior.[2] Notice that these two models underpin competing branches of political theory and jurisprudence and the human sciences more generally (see Audi, 2007). Curiously, anthropology and geography are less reliant upon a single meta-theory of human nature; more often than not, we invoke "evidence" from certain times and places to select between one theory and the other.

Many in the natural sciences see human nature as the object of research rather than an a priori assumption. Without doubt, the single most important scientific model of human nature draws its inspiration from Darwin and the principles of species evolution. By this logic, whether we are selfish or

---

[2] For example, the seventeenth- and eighteenth-century Atlantic slave trade is often invoked as evidence in favor of "model 1," whereas the sustained campaign of British religious groups to abolish the slave trade is often invoked as evidence in favor of "model 2." Recent debate about our responsibilities for others' welfare, especially in Africa and in the Middle East, work back to issues of evidence for and against competing models of human nature, as well as betraying a certain moral superiority about "us" and "them."

altruistic, whether we are competitive or cooperative, and whether we are immoral or moral, are empirical questions set against human evolution rather than tautologies or strategic moves of argument (compare Axelrod, 1984). Similarly, who we are, what we are, and what we can or cannot be are issues to be resolved by reference to our biological capacities, instincts, and limits: integrating our biological heritage with observed behavior is one of the most important scientific projects of the twenty-first century (Kahneman, 2003). Nonetheless, there is disagreement about whether competition and cooperation are traits or strategies, just as there is deep disagreement about the significance of human consciousness compared to other species (Hurley, 2008).

There is, however, broad agreement about the relevance of our biological heritage for behavior as well as for our capacity to adapt to changing circumstances. Similarly, there is broad agreement about human rationality—the logic underpinning human evolution which is presumed to be a trait rather than a strategy and is believed to be deeply embedded in our cognitive make-up rather than something that people have or do not have. It should be noted, though, that cognitive scientists have sought to show that *among* human beings there is a considerable range of cognitive ability; people are not equally adept at common tasks and functions. So, for example, being numerate including being able to calculate probability is a human trait; but people are clearly not equally adept at solving such problems (Baron, 2008). There is also broad agreement that differences between human beings on these types of issues are inevitable, but are *not* attributable to social categories such as race, gender, income, or location of residence.

The biological project on human nature is a universal project and, as such, has salutary lessons for discounting endemic conflict over the status of claimed "important" differences between people. As we share a common biological heritage, we share the same types of emotions, modes of behavior, and predilections for competition and cooperation. To think otherwise, to imagine that "others" are so profoundly different or unknowable, would be to conflate observed behavior in response to certain circumstances with a claim that people from other times and places are systematically different by virtue of their human nature. Orientalism, if a common move of political rhetoric, is nothing more substantial than folkstories of distant lands (Said, 1978). Stating the obvious may appear trivial. But, this is often lost in contemporary debate over ethnicity, religion, and geopolitics.[3]

Vital for our work on pension planning is the project initiated by Tversky and Kahneman (1974) on human decision-making under risk and uncertainty. Their

---

[3] There are, of course, those from sociobiology who use the program in a reductive manner, such that observed differences between human beings by race, gender, etc. are "explained" by reference to the imperatives of environmental survival. We do not agree with this rhetorical strategy, allying our work with others such as Lewontin (1993).

research resonates, in the first instance, with the Carnegie school of behavioralism—the insights of Herbert Simon (1983) on reason and behavior and the theory of the firm associated with Cyert and March (1963). Even so, it was ignored for many years, swamped by the momentum driving the widespread adoption of rational expectations (Lucas, 1986). It slowly gained traction for a number of reasons, not least of which was the unease of leading economic theorists about the normative bases of conventional notions of rationality and the arbitrary assumptions needed to make rational expectations work in practice as desired by its advocates (Arrow, 1986). Most importantly, behavioralism was seen to be relevant to microeconomic theories of behavior in circumstances where irreversible commitments are subject to risk and uncertainty (Winter, 1986).

All human life might be so characterized. But it is clear that Kahneman and Tversky's project was focused upon contemporary life dominated by markets and a liberal culture that favors voluntary choice and self-reliance. As such, it is easy enough to see how and why it is so important for understanding human welfare in the advanced capitalist economies of the twenty-first century.[4] Being able to judge the flux and flow of markets, being able to anticipate the actions of others, and being able to insure oneself against risk are crucial if individuals are to plan effectively for the future. All the evidence suggests, however, that these three characteristics are rarer than assumed by rational choice theorists (see Chapter 4). In fact, as the global economic crisis demonstrates, relatively few people seem to be able to cope with systemic market uncertainty. These circumstances demand a level of computational and conceptual sophistication that few have at their fingertips or even appreciate as important (Allen and Gale, 2007).

Humans may be rational but are variable in the competence and consistency of their decision-making. Kahneman and Tversky show that humans are risk averse to a fault, heavily discount the future, are opportunistic where a long-term perspective may be more conducive to achieving goals, and are inefficient users of information. Following in the wake of Simon (1978), Kahneman and Tversky show that rationality is not only bounded but systematically occluded by *innate* cognitive biases. Recognizing these traits, Gigerenzer et al. (1999) suggest that humans cope by using heuristics that can compensate for these shortcomings in task-specific circumstances.[5] Acknowledging the existence of

---

[4] Even so, it is important to acknowledge that the contemporary significance accorded to individual volition has its roots in nineteenth-century political philosophy associated with J. S. Mill among others, and indeed UK pension policy at the end of the nineteenth century and at the start of the twentieth century; see Clark and Whiteside (2003) and Whiteside (2003) for relevant overviews of the history of UK pension policy.

[5] Task-specific circumstances make serious demands on decision-making competence—well beyond agents' general intelligence and education. This is not always appreciated in the social science literature, wherein it is all about information processing rather than skill, knowledge, and substantive judgment. See Clark et al. (2006) for a specific instance and Wagner (2002) for a more general statement.

cognitive anomalies and biases is a way of explaining shortfalls in reasoning while providing a structured account of human behavior in the context of risk and uncertainty. By this logic, social scientists can accept that humans are rational while appreciating the fact that observed behavior can range widely and be heavily influenced by the environment in which they live and work. To hold otherwise would seem to deny genuine engagement in the nature and scope of human nature.

Taxonomy of Customary Behavior

In theory, at least, people make decisions in a logical manner. They begin with a set of options, assess the utility of those options against likely and desired outcomes, and choose the "best" option, assuming that they can only choose one option at any point in time (Schick, 1997). The "logic" of any action is the product of conscious deliberation. But, are people really like this? Do they make decisions in this manner? These questions animate contemporary psychology and cognitive science. There are a variety of answers, most of which seem to suggest that, when people react to situations, they utilize a variety of decision techniques, most of which do not seem to require the degree of deliberation implied by rational choice theory. For example, Gigerenzer et al. (1999: 13–14) reject optimization in favor of Simon's notion of satisficing, and suggest that people typically use "fast and frugal heuristics" because these devices economize on time and information and can be adapted to "real environments."

There are, however, significant issues to be considered, notably the mechanisms of choice among applicable heuristics, as well as what happens when the available heuristics are not relevant. Here, there are three related problems. First, the heuristics model of decision-making may be, paradoxically, a trap. By treating heuristics as a set of decision tools, deliberation may be reintroduced by the back door as the means of breaking deadlocks between the available and applicable heuristics. Second, by focusing upon the instrumental nature of heuristics, Gigerenzer and others seem to promise long-term decision effectiveness: the anomalies and biases noted in everyday decision-making are presumably slowly resolved in the face of ever finer heuristics that are applied time and again to perfection (Bermúdez, 2009: 168). Third, for heuristics to work as expected or hoped, the "background" environment must be sufficiently stable and predictable for people to have confidence in the efficacy of inherited decision tools. Stability is in short supply in financial markets.[6] An

---

[6] If heuristics are expressions of past choices honed to perfection on a case-by-case basis, this is only possible if history repeats itself. The problem is that we cannot be sure whether an event is a

alternative is to invoke "states of mind" that can do the hard work of explaining the nature and scope of behavior without returning to heuristics or conscious deliberation.

In this regard, Hogarth (2001) suggested that *intuition* is an essential component in any comprehensive taxonomy of behavior. Using synthetic examples to illustrate his argument, including one drawn from financial trading, he suggested that intuition is a form of unconscious judgment prompted by immediate circumstances and deployed without the benefits of contemplation and the steps of logic associated with the theory of rational choice. In action, the efficacy of intuition would appear to rely upon people's confidence in their ability to respond to events which in turn can be thought to rely upon the familiarity of the situation, the apparent pay-offs of one response over another, and the sense in which analogous situations were dealt with in a similar manner in the past (Tversky and Kahneman, 1974). Decision analysts believe that intuition is effective because it is immediate, it economizes on the costs of conscious deliberation, and it acknowledges the lessons found in sequential action (Kahneman, 2003).

Intuition is presumably subjective—being an expression of a person's emotional state and sensibility. Consequently, intuition could be thought entirely personal, even arbitrary, when one person is compared with another and when one person's actions are tracked over time and space. Not surprisingly, there is considerable interest in "weakness of will," a notion that could be thought to represent the ennui of life (see Ainslie, 2001). If important, the role of public policy is clear: those lacking confidence in their financial planning ought to be encouraged to develop the formal skills of decision assessment in the hope that by trial and error they become more accustomed to making decisions against reliable benchmarks. Whether this is really possible, given that psychologists believe that confidence is innate rather than learnt, given the correlation of confidence with gender, and given the lack of reliable benchmarks, is open to dispute. It does, nonetheless, underpin government policies around the world (see Strauss, 2008*b* on the UK Financial Services Authority).

But it would be strange to imagine that people are just spontaneous, signal-response machines unfazed by the future. In fact, a bedrock concept of behavioral and social psychology is *habit*. Defined by Rachlin (2000: 7) with reference to William James, it "functions in human life as a flywheel functions in a machine, to overcome temporary opposing forces, to keep us behaving for a time in a particular way, according to a predetermined pattern." In contradiction

---

repeat of something known or the start of something new (Winter, 1986). Yet, the opportunity costs of non-response can be extraordinarily high. See Clark (2008) on the scope of risk and uncertainty in financial markets.

to the theory of choice, habit is believed to be the means by which the tyranny of choice is "managed," replacing the exercise of deliberation time and again with behavioral templates that sustain decision-making over time and space (Suppes, 2003). Unlike intuition, habit can involve a significant element of social regard. Habit is not only personal, it is also relational in that others may rely upon the habits of people they know so as to anticipate their actions. Equally, habit may claim a certain cultural affectation that makes it a shared set of aspirations.[7]

Habits are like heuristics in that they provide recipes for action. But the contrast drawn here is between habits as practice and heuristics as decision tools, wherein the former is representative of patterns of unconscious behavior in certain environments, whereas the latter is representative of a method of deciding what action to take given the known options. Because habits are formed in relation to the expectations of others who rely upon their performance, our reference point here is also Harré's notion (1993) of "social action as problem solving." As such, it is an expansive notion of behavior that is related, in part, to seeking others' approval and appreciation. By this logic, habit is derivative; being part of a larger set of actions and affectations, when taken as a whole habits are the life of human beings. Evidence on how people "properly live" is developed in subsequent chapters, noting that there appears to be a premium on intimate relationships with others (see Chapter 4).

Habit provides a template for behavior over time and space. Going further, however, analysts believe that *imitation* is the means by which behavior is anchored in society and culture. Imitation is, simultaneously, innate and socially conceived. It is one of the basic mechanisms by which people (and animals) learn from others. As such, it underpins human development. If believed simple-minded, Hurley (2008) suggested that it is actually a sophisticated and difficult task to realize ambition, given the costs of adapting behavior to that which is held to be desirable and given the costs of sorting through the interaction between the environment and behavior to identify that which can be copied. We also imitate those we respect and admire. We do so even though they may live in a world quite different from us. In this sense, imitation need not be particularly effective or efficient; it commonly represents deeply held aspirations which are often valued in their own right (beyond the expected consequences of imitation).

We could argue that the locus of intuition, habit, and imitation is necessarily "local" if what we do in reaction to events and circumstances is informed

---

[7] There are many ways of illustrating these two points. Consider, for example, the back page of the "How to spend it" section of the Weekend Edition of the *Financial Times*. There guests recount their habits of weekend relaxation, often invoking their preferred retreats, what they eat and where they eat, and how much they value (as their families value) the reassuring nature of predictable pleasures in well-defined environments. See "Perfect Weekend: Camille Goutal" *Financial Times*, May 2, 2009, p. 62.

by who we are in relation to where we live and work (Camerer, 2008: 371). Recent empirical research that has shown the relevance of this scale of life for framing decision-making resonates with many human geographers (see Schwanen and Ettema, 2009). The idea that behavior is *ipso facto* "local" is also consistent with disciplinary expectations and commitments (Smith and Easterlow, 2005). We would suggest, though, that if intuition and habit are inherently "local," the spatial coordinates of imitation may be more difficult to allocate to just one spatial scale. Imitation is relational, albeit relational at a variety of embodied *and* disembodied scales. We are, in any event, skeptical of the premium attached to the "local," believing that there is a strong normative claim often embedded therein which supposes that the "local" is the "proper" scale of life against the scaling-up effects of globalization and information technologies (Couclelis, 2009). The geographical scale of intuition, habit, and imitation should be subject to the same empirical scrutiny as risk aversion, etc.

Practice of Decision-Making

Kahneman and Tversky's project is especially relevant to the contemporary world, dominated as it is by market imperatives. In this respect, the *context* of their project, and other related projects on human reasoning such as Herbert Simon's, sets the parameters for the possible scope of individual decision-making even if context is *not* a determinant of behavior. As such, behavior is embedded in the ensemble of market institutions that are, more or less, representative of the effective structure and organization of capitalist society. Here, there are obvious connections to be made with the relevant social science literature, including Polanyi (1944) and Miller (2002) who believe that context matters a great deal. Similarly, Scott's work (2008) on the economic and social foundations of modern capitalism provides useful clues as to the origins of observed differences of behavior between different types of people (by their employment and income).

Market economies are characterized by constant change, whether by growth or creative destruction. By this logic, life is calibrated against the temporal rhythms of the economy, and is reflected in social relationships and customary practice (Glennie and Thrift, 2005). Market economies also have deeply engrained and distinctive spatial configurations. From a political economy perspective, economies make places what they are, including their social relationships and institutions. Just as importantly, places make capitalist economies distinctive, as explained by political scientists and sociologists through the logic of "varieties of capitalism," and by economic geographers committed to the logic of path dependence and clusters of innovation (Peck and Theodore, 2007). Inevitably, individual decision-making under risk and uncertainty is entwined with the temporal and spatial configuration of modern economies.[8] Time and space are both constitutive and derivative of the nexus between environment and behavior.

Context stands for the macro or structural logic of decision-making setting the parameters for behavior. If an exercise in categorical distinction, this way of expressing the issue provides a roadmap for empirical analysis. Nonetheless, it is unsatisfactory for a number of reasons. First and foremost, it privileges the economic over other social and political systems—social and political life appears derivative of the economic sphere. This is not intended. Rather, it is the inevitable result of a certain emphasis on those contemporary environmental forces that structure decision-making under risk and uncertainty. As such, it is an analytical strategy rather than a historical approach to the evolution of economic, social, and political institutions that might provide for different conceptions of the proper significance of one sphere of life over others (Engelen, 2003). If it is silent about culture, there are obvious links to be made to the cultural turn in economic geography (see Scott, 2006; Scott and Power, 2004).

Consider the significance of two types of widely recognized behavioral traits. The accumulated research suggests that people have high discount rates (to the extent they have coherent discount functions), are slow to respond to changing circumstances, and generally value the near-future over long-term plans, even if this behavior is known to be to their detriment (Ainslie, 2001). Similarly, research suggests that people are poor at searching for relevant information, tend to act on the basis of immediate experience, and find probabilistic reasoning problematic. In sum, they are "myopic": they rely upon what they know even if that is specific to certain times and places (see Kahneman et al., 1997; Sedlmeier and Gigerenzer, 2001).[9] These traits suggest that, in the absence of cues to the contrary, people tend to act local: the significance of an event or opportunity is perceived through the lens of what they know and they are hostage to events that happen elsewhere and feed back over time into their immediate "environments."[10] Contrary to models

---

[8] As suggested by Langley (2008), the hegemony of finance is a distinguishing feature of twenty-first-century capitalism, where the temporal contingency of individual well-being has permeated the fabric of everyday life in Western societies. The global financial crisis has served to make this transparent, and to appreciate the scope of financialization in Western economies.

[9] It should be recognized that these traits are just two sets of a growing list of so-called biases and anomalies. Krueger and Funder (2004) identified about forty such biases, and Baron (2008), in the latest edition of his book, listed about sixty such biases. Notice that the implied order of these two sets of biases may not be stable, in the sense that the evidence suggests that people are susceptible to over- and underreaction to the environment and that governing these types of reactions may take more conscious deliberation than many people seem willing to commit.

[10] This is expressed in many different ways relevant to economic geography and financial theory. See, for example, Huberman (2000) and Wójcik (2009) on "home-bias" (the observation that portfolio managers tend to overload on local securities) and Huberman (2001) on the apparent value attributed by investors to "familiarity" (the observation that individuals tend to invest in

of recursive learning in the formation of market prices (Vives, 2008), most people are not Bayesian analysts (as in the agenda sketched by Oaksford and Chater, 2007).

That people respond to events through intuition, habit, or imitation and that the temporal and spatial scope of response is customarily local rather than global provides a datum against which to analyze behavior (as suggested by Hägerstrand's 1970 "time geography"). But, we must take care not to exaggerate the significance of these findings, suggesting directly or by implication that customary behavior exhausts the repertoire of possible behavior. Here, though, expectations break two ways. There are those, like Kahneman (2003), who believe that most people, most of the time, behave *within* the parameters set by cognition and context, trading on the cues available from each blade of the scissors to arrive at satisfactory outcomes. In this respect, and in contrast to those who hold fast to the theory of subjective expected utility maximization, satisficing is not only the "normal" response to events, it is socially acceptable in the sense that others similarly inclined set the norms of expected behavior (an explanation of herding behavior in financial markets that has a strong cultural overlay; see Shiller, 2002).

This is an invitation to take steps at least to dampen or manage the effects of "myopia" if not expunge it from human reasoning (Doherty, 2003). But it takes conscious deliberation and conceptual understanding to break with that which comes naturally and thereby transcend the local in favor of the global.[11] A number of arguments or propositions can be made to the same effect: people can be taught to reflect on their plans prior to taking action, *and* the possible pay-offs to deliberation based upon non-experiential information sources from distant sites of activity may encourage reflection prior to action (Couclelis, 2009). Embedded in this multi-factor argument are a supposition and an empirical project: the supposition is that reflection is as much about information acquisition as it is about deliberation. The empirical project is to show how the available resources including information (or lack thereof) contribute to the exercise and quality of conscious deliberation and the sense in which making a break with customary practice has short- and long-term value for those concerned.

This implies the existence of a range of responses to a common environment, partly driven by the scope of reflection as well as the spatial distribution of information resources that, in some way or another, encourage more or less

---

local companies, even overloading on the stock of the companies that employ them). See also Clark and Wójcik (2007).

[11] Recent research on the management of labor markets from the local to the global reminds us that public and private institutions can play important roles in transcending the local in favor of the global, thereby discounting the cognitive skills needed by individuals to make the leap beyond everyday life (see, generally, McDowell et al., 2008).

informed reflection. We could also invoke Hogarth's suggestion (2001) that some environments are more or less conducive to human reasoning and that a predilection in favor of reflection is mediated through the social resources needed to evaluate effectively the options for action. As well, Scott's thesis (2008) that modern capitalist societies rely upon cognition and culture for the production of value may be especially significant, providing a rationale for the internal stratification of societies wherein people occupy certain roles and responsibilities that require (or do not require) the deliberate exercise of judgment. By this logic, the scope of observed behavior—beyond intuition, habit, and imitation—is "produced" by society (the environment) and is "regulated" by the performative requirements (context) that attend certain social roles and responsibilities.[12]

To illustrate, consider the implications of two recent papers, by McDowell (2008) and McDowell et al. (2009), on the stratification of the UK labor market and the demand for certain types of job-specific "skills." In her work, as in Scott's work (2007), the labor market is usefully divided into two: at one level, there are relatively high-paid and secure jobs that require education and formal training and skills; at the other level, jobs are insecure, poorly paid, and require little in the way of formal education. These jobs require "deference and courtesy" rather than calibrated judgment and fine-tuned response to changing circumstances. In other words, the nature of work can provide "kind" environments (at the first level) that reward the exercise of judgment and deliberation in ways that reinforce the value of these skills, and "wicked" environments (at the other level) that deliberately discount the value of such skills in favor of "performance" regulated by discipline and conformity. We know little about the crossover effects of social roles and responsibilities in specific settings for the nature and scope of "private" behavior. But, the implications are plain.

For some analysts, the high level of cognitive skills needed to perform some types of work in modern economies represents a possible pathway back to reason and rationality, albeit highly stratified and segmented. So, for example, governments around the world have optimistically promoted informed decision-making among those people who have the cognitive capacity and material interest to trigger self-reflection and responsibility (in the face of the looming social costs of ageing and welfare). By contrast, Thaler and Sunstein (2008) have argued that it is better to work with customary behavior rather than against innate traits to foster desirable individual and social outcomes.

---

[12] It is shown in Clark et al. (2007) that people with task-specific expertise and domain-specific knowledge tend to be better than "neophytes" at solving investment-related problems. That is, the domain in which they live and work is an essential ingredient in how they respond to risk and uncertainty.

By this logic, institutions are best advised to design "channels" or "streams" of action in response to the environment, such that these devices prompt behavior that "automatically" transcends the myopia of cognition and context. Implicitly, perhaps, Thaler and Sunstein suggest that the discounting of cognition in many segments of the modern economy means that there is a real need for institutions that can compensate for the lack of calibrated judgment in everyday life.

Behavioral Economic Geography

The behavioral revolution has much to say about our cognitive heritage; less to say about the role of the environment in affecting behavior; and even less to say about how Simon's scissors function (i.e., the neurological interaction between cognition and context). Furthermore, the vast majority of published studies of cognitive function and performance rely upon the responses of university students (undergraduates and postgraduates) to accepted problems and puzzles (Baron, 2008). That these puzzles are replicable, have established protocols for implementation, and carry with them recognized patterns of likely outcomes provides those working in the field with a toolbag of controls and methods. Without these protocols it is doubtful that the behavioral revolution would have been anything more than a critique of the assumptions and methods of economics and the related social sciences.

However, the respondents of most studies are naïve and inexperienced. They have little or nothing at stake in their responses. This is deemed, by many, to be an advantage in that age and experience may prompt second-order disputes over the design of puzzles that are confounded by everyday life. Still, the institutional context of these experimental regimes is unsettling. At best, these respondents provide us with a baseline for testing variations in cognitive function and performance (Clark et al., 2006). At worst, utilizing these respondents in "plain vanilla" settings implies a set of troubling assumptions. First, that these respondents are representative of society; second, that social identity, roles, and responsibilities are irrelevant in assessing cognitive function and performance; and third, that education, experience, and task-specific decision routines only mediate cognitive traits. Notwithstanding the fact that experimental protocols are set so as to be relevant to risk and uncertainty, there is also little appreciation of culture-specific definitions of risk (Douglas and Wildavsky, 1983).

By raising these concerns, we do not mean to overturn what has been achieved; cognitive traits do not evaporate once we take into account culture or society (Williams, 1995: ch. 7). Rather, there remain crucial empirical issues to be resolved. For example, can age, experience, and task-specific education and training make up for apparent shortfalls in cognitive function and performance among naïve respondents? We think so, as do others, especially those convinced of the significance of learning (individual and social) (Doherty, 2003). If, as we believe, there is a natural bias in favor of that which people directly know and experience, is behavior always "local"? Are there "tricks" that can enable people to work up the geographical scale of intention and behavior? Answers to these questions have enormous implications for human development (cognitive and social), just as they have enormous implications for understanding the map of economic development and human welfare (within and between countries).

We are convinced that social roles and relationships make a difference to the consistency of decision-making under risk and uncertainty (Clark et al., 2007). This can be an argument about measured cognitive function and performance; it could also be a theoretical argument about human nature, referencing the innate sociability of human beings. One of the shortcomings of Kahneman and Tversky's research program is its exclusive focus on individual behavior (Krueger and Funder, 2004). On this point, Baumeister (2004: 15) stated that "the need to belong is a central feature of human motivation" and that emotional attachment can sustain decision-making even when the scope of risk and uncertainty is such that individuals are paralyzed by inaction. Equally, social rejection can prompt a level of emotional distress that scuttles even the simplest of decision rules. Missing from many accounts of decision-making are the social commitments and relationships that are expressed *through* intuition, habit, and imitation and *mediate* cognitive biases.

The primary research and consequent evidence for our argument is demonstrated in subsequent chapters in this book. In Chapter 3 we test whether it matters if people are young or old, male or female, and with a spouse or not for individual risk assessment. Using a set of questions designed to elicit respondents' intentions with regard to risk and uncertainty, it was found that risk predisposition and behavior were correlated with socio-demographic status and income, and that having a partner with a pension entitlement makes a difference to risk propensity, all other things being equal. It was also found that low-income, single women are highly risk averse and appear to discount any future perspective. In effect, social identity and relationships are correlated with individual risk-management strategies: some people make plans for the future based upon their intimate relationships with others (see also Friedberg and Webb, 2006). Respondents' savings intentions are shaped by their immediate relationships with others, and are therefore "local" albeit with two possible effects: positive and negative (as suggested in Chapter 1).

In Chapter 4 we begin with research about the scope of conscious deliberation. We rehearse the argument (above) about the limits of deliberation, thereby providing a taxonomy of planning—from the reactive, the responsive, to the deliberately forward thinking. Here, we are interested in who is a planner and who is not. As in Chapter 1, we are able to show that social identity, age, and income, along with intimate relationships with others, can make a significant difference to the nature and scope of deliberation.[13] We seek to determine whether people scale up their search for advice and information in making planning decisions: that is, whether people are always "local" or use geographically remote, disembodied sources of information such as the Internet in calibrating their plans. We found that local relationships dominate disembodied and institutional sources of advice and information. Our respondents rely upon those they know whether by trust or by contract (*contra* Couclelis, 2009).

Notice, though, that the interpretation of these results is open to two possible arguments. On one side of the equation, planning may reinforce social advantage and the value of existing beneficial relationships (wherein planning is an instrumental strategy). On the other side, planning may be derivative of social status, in that it refers to shared aspirations for the future reinforced by their relationships with others, rather than the tactical advantages of social status (an argument about culture and context). What is unresolved is whether people plan for the future because they can, because their peers do so, or because the available resources, including advice, match their expectations of the necessary "localness" of salience and significance.

To bring the research program to the global economic crisis, in Chapter 5 we develop a model of behavior in time and space that allows for the coexistence of different types of response to changing market conditions. From the "sophisticated" market players introduced in Chapter 4 to "naïve" and "myopic" players, our framework allows for the categorization of behavior at the intersection between cognition and context. It is apparent that some people anticipate the future, put in place so-called "safety first" strategies to manage the downside risks associated with market turmoil, and have an understanding of the world that is able to bridge the local and the global. These people are, however, unusual. Through a study of the retirement savings strategies of employees in a London-based investment firm, we show that they are of a younger age, have the advantages of income and property assets, and can distinguish between the short term and long term. From these findings, we are able to explain why, even in the most significant financial crisis since the great depression of the 1930s, not all market players were "victims" of market volatility.

---

[13] Our findings for the United Kingdom match those of Lusardi and Mitchell (2007) for the United States, wherein they show that many people do not "plan" for retirement; those who do plan tend to have higher levels of education and income, whereas those who do not plan tend to have lower income, lower levels of education, and be from ethnic minority households.

Chapter 5 focuses on the interaction between various forms of retirement saving and investment, emphasizing the role of property in savings portfolios. Because of the distinctive tax treatment of buy-to-let property, it is reasonable to suppose that this form of long-term saving complements other types of retirement saving. In Chapter 6, however, we turn to a far more complex issue: the significance attributed by our respondents to the home in which they live for retirement savings. As Smith (2009) and many others have noted, the family home is more than an investment; it is a social marker, an emotional commitment, and a resource. If financial markets and property markets are subject to risk and uncertainty, is the family home a safe haven for storing value? From our research, it would appear that older, higher-income people do believe this to be the case; further, it appears that those who would rely upon the family home for some portion of their retirement do so believing it carries risks that are more "knowable" than those embedded in financial markets. Here, we use a variety of risk metrics to test out people's risk preferences, including Kahneman and Tversky's justly famous prospect theory (1979). If our respondents are more sophisticated than most, their risk preferences are best explained using simple risk metrics and not complex representations of risk.

Chapter 7 combines a commitment to understanding the nature and scope of decision-making with the realization that changing economic circumstances, filtered through individual attributes, can make a difference to decision-making. Through an analysis of the intended take-up (or not) of annuities at different ages postretirement, it is shown that social status and identity interact with geographical location to produce highly differentiated maps of personal initiative. This study was cast in the aftermath of the 2000/1 stock market crash (post-9/11) and sought to test observed responses against continuing debate over the putative value of annuities for defined contribution pension plan participants. It was found that the UK map of private pension benefits is significantly related to social status, and that the intended take-up of an annuity is significantly related to the region of residence of respondents; there is, in effect, a UK map of risk (and non-risk) cultures, such that reliance upon financial markets systematically varies by region. As we note, the purchase of an annuity for postretirement is a very contentious issue, especially for those who might be obliged to make a compulsory purchase, either by reason of public policy or scheme rules. We show, moreover, that people's preferences change as they age, such that there are very different responses post-70 years of age.[14]

---

[14] These findings resonate with recent research by Leyshon et al. (2004, 2006) and others on the "financial ecology" or landscape of UK cities and regions as regards the production and consumption of certain types of financial products.

Going further, in Chapter 8 we ask who would exercise control over the management of their pension plans. This is an important issue and is a prelude to the final chapter of the book, which deals with the design of pension systems. The Chicago school of libertarian paternalism exemplified by Thaler and Sunstein (2003) believes that pension providers ought to take advantage of the lack of interest of people in pension planning. They would deliberately frame the options available to participants so as to dampen the effects of cognitive biases on participants' decision-making and management of their pension savings. We find that those plan participants who would claim control over crucial decisions taken with respect to their pension welfare have many of the same characteristics as long-term savers noted above, including their reliance on partners for planning effectiveness. Libertarian paternalism comes close to coercion when the design of decision-making options so channels participants' behavior as to deny their initiative. This may be justified by reference to those willing to accede to the power of the benevolent "planner." But the point here is that paternalism can coerce even if coercion is deemed in the best interests of most people.

With our empirical project, and our findings brought through to the financial crisis and the implications of the crisis for understanding the nature and scope of behavior, we turn in Chapter 9 to the implications of our research for conceptualizing behavior and the design of pension institutions. The conceptualization of behavior is, clearly, an issue dealt with time and again in the social sciences and geography. To give this discussion extra "bite" we are also concerned with the implications of our work on behavior for public policy and in particular pension policy and its institutions.

In doing so, we focus on Western economies with their various commitments to public and private planning for long-term welfare. Obviously, there are significant, long-term differences between advanced Western economies on this count (see Clark and Whiteside, 2003; Ebbinghaus, 2010 for recent accounts of those differences and their trajectories). As well, there are considerable differences between countries in their mooted "solutions" to the apparent crisis of confidence in existing institutions. We argue in Chapter 9, however, that too many "solutions" fail to take seriously the scope of behavior evident in our study and the fact that many people remain "time–space myopic" about their current circumstances and future needs, notwithstanding the attempts by government to bring them into the ambit of the rational actor model. Here, we sketch the logic for a mix of public and private institutions that take seriously the scope and scale of behavior.

Our analytical strategy assumes that context (market risk and uncertainty) and the myopia of respondents (cognition) frame respondents' intentions and the consequent decisions taken about pension and retirement planning. This is a fairly conventional view of the interaction or relationship between market and behavior. We have not been able to explain how culture affects the perception of risk and uncertainty such that intuition, habit, and imitation are the mechanisms through which those perceptions translate into distinctive space–time patterns of pension planning. While the evidence presented does suggest the existence of distinctive regional risk cultures, we also need a story about the production and reproduction of risk cultures that goes beyond commonplace assumptions that risk cultures are myopic. In effect, we need to make good on the increasing evidence to the effect that financial products and market performance are the result of context *and* behavior (Hong et al., 2004, 2005). By this logic, the analytical separation of context and cognition may be resolved in the simultaneous formation of market and intended behavior (Abolafia, 1996).

Reconceptualizing Personhood

Throughout the entire book, it is shown that savings intentions and expectations can be usefully framed through the lens of our respondents' age, gender, income, and much more. Not only does it appear that *who* we are makes a difference to planning for the future, it appears in many cases that *where* we are in time and in space also makes a difference to conceptualizing and judging market risk. Furthermore, a significant finding is that our intimate relationships are crucial for understanding who among us has confidence in planning for retirement and indeed whether planning for retirement is even possible. It is intriguing, in fact, that relying upon another person or institution for future welfare implies a significant degree of trust, such that our respondents willingly give up a sense of responsibility for planning for their retirement.

Who we are, our position in time and space, and our relationships with others must be crucial ingredients in any fully fledged theory of saving for retirement. Yet, if we look for inspiration from the social sciences for the study of retirement saving, we are often disappointed by the analytical framework if not the robustness of the empirical results. More often than not, the "object" of analysis has individuals stripped bare of identity, context, and relationships. The representative agent appears neither old nor young, neither male nor female, neither married nor single, and without children or commitments to others, whether they be past generations or future generations. It could be argued that this strategy of stripping bare the analytical object of theory building has the great advantage of isolating the crucial variables for further analysis. It could also be argued that whether a person is old or young, male or female, or of high or low income are contingent variables that only modify outcomes to a greater or lesser extent; from the representative agent to the particularities of social position becomes, by this logic, an empirical project.

This is not the place to mount a large-scale challenge to the assumptions embedded in the notion of the representative agent. In any event, whole disciplines are focused on explaining the nature and scope of human life and relationships premised upon the assumption that social identity, however constructed, is an essential building block in understanding behavior. However, it is important to acknowledge that the representative agent is not only a convenient fiction of economic theory; it is, as well, found in government policy when the focus of attention is upon facilitating financial decision-making in the light of acknowledged behavioral anomalies. We have in mind the decision trees promoted by government agencies which make few allowances for where people come from or their relationships to others. Rather, the emphasis is on convincing those who log on about the virtue of experimenting with decision trees, such that the rational path to the future is to be found in a set of simple universal decision cues. We do not doubt that decision cues have value: but we do suggest that one of the reasons why remote sources of advice about retirement planning are ignored by so many people is that they are so stripped bare of social identity and context that hardly anyone can see themselves in what is assumed by the software to be a proper planning agent.

Intriguingly, the limits of the representative agent have been recognized by those responsible for building retirement plans and designing decision cues sensitive enough to promote positive, reinforcing behavior. Once we become responsible for designing retirement systems, whether in the public or private sectors, it is almost immediately apparent that whether participants are young or old are crude but effective proxies for whether there is a premium on experience in making financial decisions. So, for example, if people are in the habit of saving and if they have in mind others whom they wish to emulate or in some sense wish to match in terms of behavior, then it may be far easier to encourage higher savings rates than would otherwise be the case. If we add in the fact that it appears that people over the age of 40 are more likely to realize the urgency of saving for the future, then we have a number of crucial "variables" for the design of pension vehicles relevant to both public policy and plan sponsors concerned about the long-term adequacy of provided pension instruments.

By this account, habit and the imitation of others intersect with social identity and in particular age, gender, income, and even household status. But notice that these same "variables" may be ineffective cues for others who neither occupy the same social position nor have the same sense of urgency. For younger people, other "variables" may have to be found if we are to promote the habit of saving for the future.

Of course, one of the great advantages of the representative agent as a reference point for empirical analysis is that we need much less in the way of empirical observations than if we take seriously social stratification and context in time and space. Indeed, one of the hard lessons learnt in our research was that by the time we selected people of a certain age, gender, household status, and income for close analysis as regards their risk propensities and asset allocation predispositions, it became obvious that we simply did not have enough degrees of freedom in terms of the total volume of people in our representative sample to estimate models capable of discriminating respondent behavior on the basis of their shared attributes. This becomes obvious in Chapter 3, where we look closely at issues of risk tolerance and investment. So, it is reasonable to warn the reader that notwithstanding our commitment to understanding people's savings intentions by virtue of who they are, where they are, and what their relationships are with others, we do so only up to the limits imposed by the material we have to hand.

Perhaps this is inevitable. But there is a valuable lesson embedded therein: if we want to understand the nature and scope of behavior, we need richer resources than those currently available and those that animate debate over the interaction between environment and behavior.

APPENDIX
Nomenclature

Cognition is "the act of thinking" (Hogarth, 2001: 15–16). This statement is intended to be descriptive rather than prescriptive. Hogarth suggests that most thinking is neither conscious nor deliberate. If we assume, for the moment, that thinking is analogous to information processing, cognitive scientists believe that processing is typically automatic and reliant upon mechanisms that "are hardwired in the sense that they are part of our genetic inheritance." Because information-processing routines are assumed to be second nature, it is contended by experimental scientists that these routines are based upon "experiential learning and, as such, (are) specific to particular domains" or environments (p. 23). Therefore, learning need not be effective or even possible (see Gertler, 2001 on the limits of learning between jurisdictions).

The *environment* is normally described in physical terms: that is, the environment is the world around us, where we act, what we are affected by and, in turn, that which is affected by our actions. More often treated as a concept than a "real" place, in psychology it is assumed to have the resources needed for humans to survive, if not always prosper (see Gigerenzer et al., 1999: 13). Hogarth (2001: ch. 3) suggested a taxonomy of environments referring to "kind" and "wicked" environments, wherein the former are favorable to learning while the latter are not. See also Smith and Easterlow (2005) on "(Risky) Spaces" and "(Healthy) Spaces" and the limits of environment-centered explanations of human morphologies. For all the significance attributed to the environment in human behavior, it is often treated as a passive element in human life. As an exception, see Astuti and Harris (2008) for an imaginative analysis of the interaction between cognition and the environment, showing that children's conceptual development is dependent, in part, on their tactile engagement with the environment.

Context is typically used in two related ways. In many respects, it is simply shorthand for the environment (as implied by Smith and Eaterlow, 2005). In some cases, though, context has a narrower meaning referring to the specific domain in which behavior takes place, including agents' roles and responsibilities as well as the social commitments related to that domain (Hogarth, 2001: 230–1). By this interpretation, context is shorthand for the frame or parameters that set the probable scope of individual behavior.[15] See the volume edited by Pettit and McDowell (1986) on the philosophical status of context in relation to behavior. Glennie and Thrift (2009: 395–7) use the concept to explain what prompts innovation, arguing that analysts who emphasize the initiative of lone inventors miss the crucial significance of "regional and local networks and temporalities" and, in particular, "communities of practice." See also Storper (2009).

---

[15] The notion of framing is also utilized, with variations, in social theory as in anthropology (e.g., Goffman, 1974), the study of social control (Foucault, 1977, 1979), and market systems (Callon, 1998). This connection is explored in Strauss (2008*b*, 2009*b*).

3

Risk Propensities

The pension systems of most countries are complex, difficult to understand, and crucial for the retirement income of the elderly and the baby boom generation. The UK Basic State Pension (BSP) is the largest component of the system, and many people rely upon this pillar for their well-being. For a significant minority (the middle class and those in public sector unionized jobs), however, this source of retirement income is supplemented by the second tier of occupational pensions and/or the third tier of private pension savings. On average, UK retirees derive about 60 percent of their income from state sources and about 40 percent from occupational and private pensions (Department for Work and Pension [DWP], 1998). The previous Labour government expressed the desire to control future social security spending (to 2050) by reversing this ratio such that future cohorts will be, on average, reliant on private savings and occupational pensions for about 60 percent of their postretirement income. It seems highly unlikely that the recently elected Conservative government will do otherwise, given the fiscal imperatives imposed by the global financial crisis.

The real value of the BSP has been falling for many years.[1] Furthermore, it appears that many workers (especially women and low-income earners) are not saving enough to ensure an adequate retirement standard of living. In response, the Pensions Commission recommended establishment of the National Pensions Savings Scheme (NPSS) along defined contribution (DC) lines with modest tax-preferred contributions from both employers and employees. The NPSS is intended to boost future retirement savings, especially among those not covered by private pensions, and address problems of equitability among different segments of the population including women. The NPSS will provide limited investment choices and a default option for those

---

[1] Following the recommendations of the Pensions Commission's report (2005), the government has, however, pledged to re-establish (by 2012) the earnings link severed by Thatcher's Conservative government. The basic state pension is currently indexed to prices.

unwilling or unable to make informed choices (see the DWP's 2006 White Paper on the design and structure of the NPSS).

Based on a specially commissioned 2006 survey of nearly 900 randomly selected British occupational plan participants, this chapter and Chapters 4 and 7 focus on the relationship between respondents' socio-demographic characteristics—in this case age, gender, household income, marital status, and spousal pension status—and their risk propensities. Our goal is to understand better the significance of socio-demographic factors for occupational pension plan members' risk preferences, and how those preferences may translate into intended asset allocation strategies. This is important given the current debate about the pension crisis and policy emphasis on pension flexibility, individual choice, and responsibility. Here, we seek a rapprochement between those who focus upon cognitive skills and those who believe that individual attitudes and behavior are best understood at the intersection between cognition and the environment including social structure, social identity, social values, and opportunities (see Krueger and Funder, 2004; Strauss, 2008*b*).

In the next section, the relevant literature is discussed. The structure of our sample survey and the methodology underpinning the analysis are then set out, where we point to evidence which shows that respondents tend not to differentiate risk either according to the means of retirement saving or the type of pension plan. Thereafter, the results of the analysis are elaborated, noting the significance of social differentiation and the relevance of spousal pension entitlements for respondents' risk-related attitudes. In the penultimate section, we show how respondents' social status may give rise to distinctive asset allocation strategies, suggesting that a deeper understanding of these patterns should be the subject of further research. The chapter concludes with a summary of the results and their implications for public policy, as well as the design of employer-sponsored pension education and information-awareness programs. Note that we have resisted the temptation to reproduce the tables and statistical results found in the paper upon which this chapter is based. Readers interested in the statistical details should consult Clark and Strauss (2008).

The Behavioral Revolution

Global competition, changes in accounting rules, and the stock market pricing of expected liabilities have undercut the viability of private defined benefit (DB) pension plans in much of the Anglo-American world (Clark and Monk, 2007). While the growth of private DC pension plans has been attributed to plan sponsors' lower risks and costs, DC pensions are also more flexible and portable than DB pensions—characteristics that have made these types of plans attractive to younger workers (Clark and Monk, 2008). Even so, when

Risk Propensities

plan sponsors shift from DB to DC plans, employer and employee contribution rates tend to fall, discounting the prospective relative value of such pensions (Munnell, 2006).[2] Furthermore, and notwithstanding the virtues of immediate vesting and the portability of accumulated assets, DC pensions entail the realignment of risk: from the employer to the employee.[3] The nature and scope of individual responsibility also varies by plan. But in many cases, workers become responsible for enrolment, for their contribution rates, their allocation of assets to investment classes, and their choice of investment products and providers.

As noted in Chapter 2, we could assume that pension plan participants behave rationally and maximize utility. Yet, as Kahneman and Tversky (1979) have shown, individuals do not always act according to the dictates of economic theory, especially under conditions of risk and uncertainty. Recent research confirms that many DC participants do not make "optimal" decisions about saving for retirement (see Iyengar et al., 2004; Mitchell and Utkus, 2004*a*; Scholz et al., 2006). Some commentators have expressed doubts about the ability of plan participants to make decisions that are consistent with their long-term financial needs, citing a lack of financial knowledge and understanding (see Dolvin and Templeton, 2006; Lusardi and Mitchell, 2005). It has also been observed that employees often do not take full advantage of employer matching contributions (see Choi et al., 2002, 2004, 2005); they have less than optimal financial knowledge and do not utilize information efficiently (Bernheim, 1998); and they are risk averse when faced with asset allocation and portfolio choices but may be risk-seeking in the face of losses (see Kahneman et al., 1997; Tversky and Kahneman, 1991).

The behavioral revolution has had far-reaching consequences for pension research. But it should be recognized that individual planning and decision-making occurs at the interface between cognition and the environment such that observed outcomes depend upon many other factors including socio-demographic characteristics (Clark and Strauss, 2008; Strauss, 2008*b*, 2009*b*; Weber et al., 2002). Gigerenzer (2004) invokes Simon's scissors metaphor (1956) to suggest that a person's place in society, and by implication their relationships with others, may play a significant role in observed risk

---

[2] See also Munnell et al. (2006) showing that DC plan participants are often adversely affected by the relatively poor performance of their investment funds compared to DB plans.

[3] In DB plans the employer is responsible for ensuring that the plan is adequately funded such that an agreed level of benefit, often derived from a formula based on final salary or an average of several best years plus years of service, is paid for the life of the retiree (and beyond, in the case of plans with survivor benefits attached). With DC plans the amount of benefit is based on accumulated contributions and investment performance: there is no guarantee that DC assets will provide a guaranteed income stream in retirement unless an annuity is purchased. Although annuity purchase is mandatory after 75 in the United Kingdom, the cost of these products in relation to the average accumulated DC pension "pot" means that the de facto salary replacement rate is likely to be significantly lower than in a DB scheme.

preferences. It has been shown that differences in risk-related attitudes and behavior emerge when socio-demographic variables such as gender, occupation, and age are considered. It has been also observed that, in general, women are more risk averse than men, the young are more risk-seeking than the old, wealthier individuals manifest a greater willingness to invest in equities, and the poor are risk averse (see Bajtelsmit, 2006; Bajtelsmit et al., 1999; Bernasek and Shwiff, 2001; Ginn and Arber, 1991; Lusardi and Mitchell, 2005).[4]

Here, we report the results of tests for the significance of such characteristics for UK pension-related attitudes and planning behavior. We also show that a more complex picture emerges when these characteristics are examined in *interaction* with one another. In a study of household bargaining, Friedberg and Webb (2006) indicated that risk tolerance in financial decision-making is associated with the age of an older spouse, and whether that individual is male or female. Similarly, Hallahan et al. (2004) demonstrated for a large sample of Australians that the household, not just the individual, may be an important risk-related decision unit. In UK society, income, education, age, gender, and marital status are thought to be closely related and are proxies for financial literacy and planning. Recent research also suggests that cultural context may be very important in driving risk-related preferences *whatever* our shared cognitive biases and social institutions (see Bertrand et al., 2005 on financial decision-making and Henrich et al., 2005 on risk behavior in a cross-cultural context).

By our assessment, understanding pension-related planning and behavior requires, at one level, an appreciation of cognitive competence and, at another level, sensitivity to the social status of respondents as well as their cultural milieu. This chapter tests for the significance of the social characteristics of UK respondents in pension attitudes and behavior, utilizing variables identified as significant in pension research conducted in other countries (e.g., on 401(k) pension plan members in the United States). As such, it is one of just a few attempts in the United Kingdom to test for the relationship between individual pension-risk propensities and their social characteristics. Nonetheless, it should be recognized that the scope of our findings and conclusions are limited by the size of the sample and our emphasis on perceived risk rather than actual behavior.[5]

---

[4] Recent work has focused upon the "ecology" of financial literacy including reference to the vulnerability of UK inner-city residents to the aggressive sales techniques by the vendors of financial products (see Leyshon et al., 2004, 2006). Underpinning this work is an argument to the effect that individual cognitive capacity combined with the environment in which decisions are made creates an uneven landscape of financial competence (see Chapter 6).

[5] There is, of course, widespread debate in the academic literature on the relationship between attitudes and behavior (see generally Hollis, 1996). Some social scientists are skeptical of self-reported risk attitudes, especially if the issues posed are not embedded in a specific domain or context. By contrast, we follow the lead of Dorn and Huberman (2005) and Weber et al. (2002) in explicitly relating respondents' risk preferences to a specific set of pension-related choices.

Survey Method and Implementation

The questionnaire survey of UK pension plan participants' risk-related attitudes and intentions was sponsored by Mercer Human Resource Consulting as part of their developing DC consulting practice. It was conducted between January and May of 2006 by the UK market research consultancy Objective Research. As such it came more than a year before the peak of the bubble, and then the global financial crisis. The original target, established by Mercer and Objective Research, was a representative sample size of $N = 1,000$ respondents to the postal questionnaire; in other words, the aim was to gather 1,000 valid survey responses for analysis.

The survey was designed to represent the population of UK workers employed by organizations with more than twenty-five employees worldwide. In addition, respondents were initially screened by the following criteria: they had some form of occupational or private pension plans; they worked in the private sector; and their pension could not be a final salary pension. In the first six weeks of fieldwork, however, it proved difficult to recruit respondents using these criteria.[6] It was agreed that no more than 20 percent of the sample should fall below the initial criteria and that the survey would recruit those working for organizations with more than ten employees, include public sector workers provided they had a private pension, and include final salary (DB) pension holders. In the end, 16 percent of the sample fell below the initial criteria of being an employee in the private sector and 15 percent of the sample also included members of a final salary pension scheme.

The sample was sourced primarily from consumer lists ordered over three batches.[7] The first batch was split into two selections: 5,000 working individuals sampled as having a company pension, split by age and gender, excluding those who were retired, at home as housewives, in public services, the self-employed, students, and teachers/lecturers; this selection was then topped up with 2,000

---

[6] It is important to recognize that, unlike the United States, the growth of DC pension schemes has been relatively slow and only significant over the last five years as traditional DB plans have been closed to new entrants or fully terminated (Pensions Commission, 2004b). By contrast, a similar survey in the United States would easily meet the selection criteria. The relative importance of DB and DC schemes in the United Kingdom and the more general implications of the slow growth in non-DB plans for private pension coverage are explored in Clark (2006a).

[7] Customer lists were purchased from an independent list broker. The lists were drawn from large consumer databases in order to ensure that the sample was composed of individuals with the types of pensions of relevance to the analysis. Objective Research specified the number of contacts required by age-within-gender, which the supplier was then asked to select on a random basis. Objective Research conducted checks upon receipt of the sample and during the survey process in order to ensure that it included a representative spread (e.g., of geographical regions). These checks were repeated during the analysis phase. The telephone screening and questionnaire editing procedures were used to verify the selection information used by the list provider and to decide whether people were actually relevant for the survey.

working individuals, in relevant occupations, who had not indicated that they held a company pension.

Objective Research conducted screening telephone interviews using home telephone numbers over a period of ten weeks from January 16, 2006 until March 28, 2006. In the third week, another batch of 5,000 was ordered, followed by another in the ninth week. Each was split by age and gender according to the quotas that the interviewers had achieved at that point. Objective Research ordered a total of 17,000 records with the aim of recruiting 3,000 individuals to participate in the postal questionnaire. The sample contained a large number of invalid and wrong numbers (over 20 percent); therefore, two top-up batches from consumer records were ordered to compensate. At the point of the second top-up order, a 10 percent error margin was agreed with the sample supplier.

From the sample of 17,000, a total of 2,151 individuals were sent the eight-page questionnaire with a cover letter (available from the authors). This was 849 short of the figure of 3,000 individuals that was agreed at the outset; due to the difficulty of finding eligible respondents, it was decided that efforts should focus on achieving a high (45–50 percent) response rate with the goal of generating 900–950 eligible questionnaires. To this end, follow-up telephone calls were made to identify respondents in addition to reminder mailings in order to boost response to the mail-out. Despite an increased volume of calls to 18–29-year-olds in the last five weeks of the survey, the response rate was particularly low among this age group, even though they were offered the incentive of a £5 HMV gift card. An overall mail response rate of 44 percent was achieved. Returned questionnaires were checked for completeness and eligibility: 99.8 percent of returned questionnaires met the eligibility criteria, giving an "eligible" response rate of 44 percent.[8]

Attitudes to Risk

The survey allowed us to examine: (*a*) whether respondents distinguish between the means of retirement saving and the type of pension plan in relation to their risk propensity; (*b*) whether respondents' expressed risk propensities are correlated with their socio-demographic characteristics including

---

[8] All replies were punch verified (i.e., the data were entered twice). Objective Research produced four sets of tables in Microsoft Word, two of which were base "All Answered," as well as a weighted data file in Microsoft Excel. The tables were weighted to the following criterion: age-within-sex based on the total UK working population statistics. At this point, the sector demographics were checked for any bias. None was found, therefore additional weighting was deemed to be unnecessary.

gender, age, marital status, and income level; and (c) whether respondents' social status can be shown to be related to desired asset allocation strategies.

Gauging attitudes toward financial risk is notoriously difficult, not least because risk is reckoned to be poorly understood by many members of the public at large when the issues posed lack an appropriate level of domain-specificity. In this survey, a specific pension-risk question was set using non-technical language so as to classify respondents by their risk propensity. Question 25 asked: *"When you are thinking about long-term savings and pensions, which of the following best summarises your attitude: (A) I aim to get the best possible growth in the value of my savings, even if that means taking some risks which could cause my savings to fall in value; or (B) I prefer to have safe and secure savings and investments, even if that means they do not grow in value as much as they could."* This question is based on Kahneman and Tversky's well-known test (1979) of risk aversion under risk and uncertainty which has been widely applied in economics and psychology (see Baron, 2008) and is the basis of Clark et al.'s analysis (2006) of the risk preferences and decision-making of UK pension plan trustees.

So as to compare respondents' risk-related attitudes by the means of retirement saving and the type of pension plan, respondents were classified as follows: 145 (16 percent) of respondents indicated that they had a personal pension, 663 (73 percent) indicated that they participated in an employer-sponsored pension plan, and 103 (11 percent) indicated that they had both. Of the 789 respondents with an occupational pension, 118 (15 percent) indicated that they participated in a DB pension, 336 (42 percent) indicated they participated in a DC pension, and 338 (43 percent) were not sure of the type of pension plan. The number of pension plan members shown to be unsure of their pension type was a significant finding in its own right. It also challenges the assumption that pension plan members make informed choices about their pension savings based on complete information and individual utility functions.

Of the 900 completed and valid responses, 202 or 22 percent chose Option A (a proxy for financial risk tolerance) and 698 or 78 percent chose Option B (a proxy for financial risk aversion).[9] This result seems to indicate significant levels of risk aversion among our sample population, but reveals little about which groups are likely to be more or less risk averse in relation to retirement income. This issue is of considerable importance for whether the estimated models ought to be partitioned. We found that no statistically significant differences exist between respondents' expressed risk propensities by the

---

[9] In this paper, "risk propensity" is used to describe an individual's willingness to assume financial risk in the context of their expressed attitudes recorded in responses to Q25. Risk aversion indicates an individual's preference for "safe and secure savings." Risk tolerance is associated with "the best possible growth in the value of my savings."

means of retirement savings and the type of employer-sponsored pension plan.[10]

Using regression techniques, we sought to determine whether there is a statistical difference in the expressed risk propensities (Options A or B) of plan members by the means of retirement savings and the type of employer-sponsored pension plan. Risk propensity was regressed against dummies representing the means of retirement savings (personal plan, employer-sponsored plan, or both) and the type of pension plan (DB, DC, or both), and tests conducted to determine whether the estimated parameters were statistically different. We first conducted paired t-tests on the mean risk propensity of those with an employer pension plan, those with a personal pension, and those with both types of pension. These tests did not find any statistically significant differences across these groups. In other words, we could not reject the null hypothesis that the mean risk propensity is the same for those with a personal pension in comparison to those with an employer pension, for those with a personal pension versus those with both types of pension, and for those with an employer pension versus those with both types of pension.

We then carried out t-tests on the difference of mean risk propensity by pairs to determine whether risk propensity varies by type of employer-sponsored pension plan. In this case, the t-tests were such that we could not reject the hypotheses that mean risk propensity was the same for respondents with a DB entitlement, compared with a DC entitlement, or for members of a DB plan compared with those not sure about their pension type. Again, all hypotheses are evaluated assuming a level of confidence of 99 percent. The test did, however, reject the hypothesis that mean risk propensity is the same for those with a DC plan and those not sure of their pension type.

Finally, we performed a Hotelling's T-squared generalized means test to confirm whether risk propensity varies statistically across groups with different types of pension (personal, employer, or both) in the sample. In this test the auxiliary variables R'_I are defined for each group as follows: $R'_I = R - \mu I$, where R is the risk-related attitude (the choice of A or B in Question 25), and μI is the risk propensity of group I. The null hypothesis is that $R'_I = 0$ for every group I. Under the null hypothesis, then, mean risk propensity is the same for every group: the Hotelling test could not reject this hypothesis at a level of 1 percent. We conducted the same test based on pension type DB, DC, or not

---

[10] We also ran the logit, probit, and linear probability models for each of the groups defined by pension type (DB, DC, Don't Know). The results of the non-partitioned model, reported in the next section, did not differ significantly from the results of the individual models: all significant coefficients kept their sign with the exception that spousal pension was positive for the DB group (however, this coefficient was not significant). The levels of significance did vary but we believe that this is a result of the models being estimated with significantly fewer observations (and therefore less robust), in the sense that the sample was not designed to be representative when subdivided by pension type.

sure and the hypothesis that risk propensity does not differ on this basis could not be rejected either at the same level of confidence. We therefore concluded that despite the results of the *t*-test for the difference in risk propensity between DC plan participants and those unsure of their pension type, we could not assume a statistically significant difference between these groups.

Risk Propensity and Socio-demographic Characteristics

The crucial finding was that respondents' risk propensity did not vary in a statistically significant way depending on whether they had DB and DC pension plan entitlements. If this finding represented the UK population at large, it could have far-reaching implications for public policy and for occupational plan sponsor responsibilities—detailed in the conclusion. More immediately, in the second stage of analysis we set out to test the relationship between respondents' risk-related pension attitudes and their socio-demographic characteristics. Gender, marital status (married or single/widowed), income, age, and a dummy variable indicating the presence of a spousal pension entitlement (yes/no) were used as explanatory (independent) variables (descriptions are provided in Table 3.1).[11] The dependent variable RISK took a value equal to

Table 3.1. Socio-demographic characteristics of the analysis sample

Variable and categories	Weighted frequency	Percentage	Variable and categories	Weighted frequency	Percentage
Risk aversion			*Gender*		
Not risk averse	201.2	23.4	Male	560.1	65.1
Risk averse	659.8	76.6	Female	300.9	35.0
Total	861.0	100.0	Total	861.0	100.0
Age			*Income*		
Young	445.8	51.8	Low	177.4	20.6
Mature	397.8	46.2	Medium	543.0	63.1
Elder	17.6	2.1	High	140.6	16.3
Total	861.0	100.0	Total	861.0	100.0
Marital status			*Spouse has pension*		
Married	659.4	76.6	Yes	433.2	50.3
Single/widowed	201.6	23.4	No	427.8	49.7
Total	861.0	100.0	Total	861.0	100.0

Source: YouGov commissioned survey, authors' calculations.

---

[11] Note that respondent yearly income was classified as "low" (<£15K), "medium" (£14K–£40K), and "high" (>£40K). Respondent age was classified as "young" (22–39 years), "mature" (40–59 years), and "elder" (over 59 years).

"1" when a person preferred "safe and secure savings" (risk aversion), and a value equal to "0" when the person aimed to "get the best possible growth" (risk tolerance).

A total of seventy-one respondents were found to have missing values in one or more of the explanatory or dependent variables and were dropped from the analysis. The final sample had 861 observations. Both the basic statistics and the econometric models were estimated using the analytical weights provided in the dataset. Three different versions of the model were estimated using logit and probit nonlinear probability models and a linear probability model. For the estimation of the risk propensity model in this chapter, the normality assumption of the disturbances is a natural choice. Therefore, the analysis focuses on the results of the probit model. However, in order to test the robustness of the results, we also estimate a logistic and a linear probability model, despite the acknowledged weaknesses of the latter.[12] The three models have equivalent results, illustrating the robustness of the conclusions and that the data do not exhibit any unusual features that hinder the estimation of a particular coefficient.

The nonlinear models were estimated under maximum likelihood (ML), and the linear probability model was calculated using an ordinary least squares regression of RISK as the dependent variable and the other determinants as explanatory variables. The probit/logit model utilized an explanatory variable y (in our case the variable RISK) that took values of 1 and 0. It was proposed that the values of y depended on the values taken by the x variables (socio-demographic characteristics). To specify a functional form of these probabilities, the probit model assumed the existence of a *latent* variable $y^*[i]$. This variable took values between 0 and 1 and followed a normal distribution. Its mean was $x\beta$, where x are the dependent variables and β the effects of the dependent x variables and its variance σ^2.

---

[12] Linear probability, logit, and probit models are estimated when the dependent variable in a regression model is a dummy variable. The linear probability model is a least squares estimation while the probit and the logit model are nonlinear. The statistical literature recognizes the limits to linear probability models: historically they were widely used as a feasible computational alternative to nonlinear models such as logit and probit, which were prohibitively expensive to estimate (Caudill, 1988). It is still a relatively common practice to use linear probability models as an easy way to obtain quick estimates in a preliminary stage of a study (Amemiya, 1981), and despite their limitations, linear probability models can still be useful to detect possible problems in data. On the other hand, probit and logit models provide efficient estimates. The difference between these models lies with the assumed distribution function: the probit model assumes a normal distribution, while the logit assumes a logistic distribution. Both models yield estimated choice probabilities that differ by less than 0.02, which can be distinguished only in very large samples. Therefore, in practice the choice between them is usually done around practical concerns such as the availability and flexibility of computer programs (Adlrich and Nelson, 1983). For a complete discussion, see Adlrich and Nelson (1983) and Caudill (1988).

Estimation of the model designed to test the relationship between the socio-demographic variables and individuals' risk propensity gave the following results:

- *Age (omitted variable, young)*: As mentioned above, young men have been shown to be more willing to take financial risks (Clark et al., 2006), and that risk propensity decreases with age (Milligan, 2004). But it is possible that this "effect" may be offset by wealth accumulation. The results for age show that the coefficient on the mature cohort was positive and significant. The coefficient for the elder cohort was negative but not significant. This suggests that there is no statistically significant difference in the risk propensities of young and old people; however, the mature cohort tends to be more risk averse. It is possible that there are life-cycle effects embedded in the responses of our respondents even if the current sample is not sufficient to disentangle the nature of those effects across a more finely distinguished set of life cycle stages. These issues are dealt with in subsequent chapters.

- *Income (omitted variable, low income)*: Recent studies of pension plan participants have shown that those who hold more assets are more likely to be less risk averse when making decisions about long-term savings (Carroll, 2000; see also Poterba et al., 2006). In our survey, both high- and medium-income people presented a negative coefficient, even if the coefficient on medium income was not significant. This suggests that medium- and low-income people present the same level of risk aversion, but that high-income people tend to be less risk averse. This result raises a question about the level at which risk aversion is neutralized by economic security: in other words, how wealthy do pension plan participants have to be in order to display, on average, a willingness to take greater financial risk? This question is broad in scope and bears upon unresolved debate in the decision sciences over the degree to which economic well-being does or does not affect risk-related decision-making (see Kahneman, 2003).

- *Gender (omitted variable, female)*: Gender differences in financial risk tolerance have been observed in experimental settings (Clark et al., 2007) and in survey-based research (Bajtelsmit et al., 1999; Papke, 1998). Our results indicate that being male has a negative effect on the probability of being risk averse, confirming the generally held view that men are more risk tolerant than women. As shown elsewhere, gender is a significant factor in discriminating between respondents' risk propensities; here, we have been able to develop UK evidence for an issue that has been widely reported elsewhere. It is important to note, however, that a sizable majority of all respondents displayed risk aversion in the context of the risk proxy (Question 25).

- *Marital status (omitted variable, single/widow)*: The effects of marital status on risk tolerance are difficult to gauge, since cohabitation with a partner suggests complex interpersonal relationships that might or might not involve pooled risk over time.[13] In the analysis conducted for this chapter the coefficient is positive, but not significant. This suggests that risk aversion is the same between married and non-married persons, or at least that marital status alone cannot account for differences in risk propensity between those with partners and those without. As shown below, it is likely that an entitlement to, or lack of, a spousal pension entitlement is more significant than straightforward marital status or gender for that matter (compare with Sundén and Surette, 1998). It would be interesting to assess whether the existence of offspring and, presumably, a bequest motive is significant in this context.

- *Spouse's pension (omitted variable, no spouse's pension)*: The coefficient for this variable was negative and significant, indicating that those respondents whose spouses also have pension entitlements tend to be less risk averse. The result suggests the interplay of two significant factors at the level of the household on individual risk propensities: higher levels of long-term savings and the pooling of risk between partners such that enhancing the total welfare of the household is the operative goal. This result is not, however, robust enough to show *how* risk is pooled within couples with different types and levels of spousal pensions (see Bajtelsmit et al., 1999 on marriage and gender differences in DC pensions decisions) or whether the effect of increased household retirement savings is the same as the individual wealth effect mentioned above.

Marginal Effects on Risk Aversion

The coefficients on the probit and the logit models do not necessarily have a straightforward interpretation since they do not represent the marginal contribution of the variable to the probability of being risk averse. This is not the case for the linear probability model. In this type of model, the estimated coefficients can be interpreted directly as marginal probabilities. In Table 6 of Clark and Strauss (2008), we presented the estimated marginal probabilities for the probit regression. It is clear from those results that the estimated

---

[13] Data on women's pension assets are hard to come by as they are often aggregated at the level of the household (see, e.g., OECD, 2004). Feminists argue that household data can be misleading; power dynamics within couples and families mean that it cannot be assumed that resources are pooled equitably. Interestingly, our results suggest that where both spouses have occupational pensions, pooling may indeed occur such that risk is shared.

marginal probabilities were almost identical in the probit and in the linear regression (see Table 3.2).

These estimations showed that having high income has the larger effect (more than −15 percent) on risk aversion. Being in the mature age cohort has a positive marginal effect of 10 percent on this probability. Being male or being married to someone with pension entitlement has a negative 7 percent effect on the risk aversion probability. The other variables did not have statistically significant effects.

We also estimated a measure of risk aversion for a "typical" agent using the results of the probit model. These results suggested that the marginal probability of risk aversion is negatively influenced by the existence of a spousal pension entitlement for most groups, with the marked exception of low-income women. The marginal probability of risk aversion is positive for most groups without a spousal pension, apart from older, middle-income men and those with high

Table 3.2. Marginal probabilities on risk aversion in the probit model

	Males		Females	
	Single	Married	Single	Married
	A	B	C	D
Spouse has pension				
Low income				
1. Young	−5.75	−1.80	1.83	5.78
2. Middle-aged	4.53	8.48	12.11	16.06
3. Elder	−7.13	−3.18	0.45	4.40
Medium income				
4. Young	−10.50	−6.56	−2.92	1.03
5. Middle-aged	−0.23	3.72	7.35	11.30
6. Elder				
High income				
7. Young	−20.99	−17.04	−13.41	−9.46
8. Middle-aged	−10.72	−6.77	−3.14	0.81
9. Elder	−22.38	−18.43	−14.80	−10.85
Spouse does not have pension				
Low income				
10. Young	1.63	5.57	9.21	13.15
11. Middle-aged	11.90	15.85	19.48	23.43
12. Elder	0.24	4.19	7.82	11.77
Medium income				
13. Young	−3.13	0.82	4.45	8.40
14. Middle-aged	7.14	11.09	14.72	18.67
15. Elder	−4.52	−0.57	3.06	7.01
High income				
16. Young	−13.62	−9.67	−6.04	−2.09
17. Middle-aged	−3.35	0.60	4.24	8.18
18. Elder	−15.01	−11.06	−7.43	−3.48

Note: The comparisons are between the average case in a subgroup and the base case, the average of all agents.

incomes (men and women) in the young and older groups. In other words, having a spousal pension increases the marginal probability that an individual will make riskier asset allocations (i.e., allocate a larger amount to equities) with her long-term savings *unless* she is a low-income woman. Conversely, being in a household where the spouse does not have a pension entitlement decreases this likelihood, except for high income and older groups, while high-income older women with and without spousal pension entitlements have, statistically, a negative marginal probability of being risk averse.

The socio-demographic profile of our respondents was such that young, low-income men who have occupational pensions were about as likely to have a wife with a pension entitlement as not. But young, married, low-income women with a pension entitlement were twice as likely to have a spouse with a pension entitlement. Middle-aged, middle-income men with an occupational pension were quite likely to have a spouse without a pension, but middle-aged, middle-income women with an occupational pension are unlikely to live in a household without a spousal pension. The same holds true for married, middle-aged, high-income men and women with occupational pensions: men were slightly more likely to have a spouse with a pension entitlement than not, but women were far more likely to do so. Only in the case of young, medium-income, married men and women with occupational pensions are men more likely than women to have a spouse with a pension entitlement. It is true, however, that a large number of these men have spouses with no pension entitlement relative to the total number of men in households with spousal pensions.

Social Status and Asset Allocation

These results suggest the existence of distinct pension-related risk "groups" in UK society, distinguished according to their socio-demographic characteristics. This is apparent in the calibrated effects of certain socio-demographic variables on the risk propensity of respondents, as well as in the interaction effects of certain related socio-demographic variables on putative asset allocation. We conclude the discussion by mapping the risk profiles of certain types of people in relation to their asset allocation strategies. Calibrating the intersection between social status, risk propensity, and actual asset allocation is left for another time (and a larger sample with questions focused on intended and actual asset allocation in DC schemes). Nevertheless, we feel that this preliminary analysis gives an indication of the relevance of the issue for further research.

In the third section of the survey (after general, personal, and demographic questions about their pension entitlements), participants were asked for their views on investment. They were first asked to rate their knowledge and understanding of three different investment fund types: cash, bonds, and equities. After providing their rating, they were given brief definitions of each type of fund. They were then asked in question 28 to make asset allocations to some or all of the identified fund types. Question 28 read: *"Imagine that you now have to invest your long-term savings for the majority of years between now and your retirement. Please indicate how you would invest your money in the three types of funds listed. Assume you have £100 to invest—how would you split between these funds."* A preliminary analysis of gender differences between asset allocation strategies showed that women allocated a mean value of £39 to the cash option, £28 to the bond option, and £29 to the equity option; whereas men allocated a mean value of £35 to the cash option, £28 to the bond option, and £35 to the equity option.

Clearly, the problem set in the survey was highly stylized given the small value of the "assets" to be allocated, and we are not able to assess the stability of allocations over time.[14] Furthermore, it is apparent that there are "missing" groups from the data: the sample did not include any people who were older, high income, and married without a spousal pension entitlement. Nonetheless, tests of association suggested that there is a correlation between risk propensity in allocation decisions and the socio-demographic variables of gender and income. To start, we transformed the asset allocation variable Q28 into a categorical variable scaled from 1 to 5 where 1 represents a very low risk strategy, 2 a low risk strategy, 3 a moderate risk strategy, 4 a high risk strategy, and 5 a very high risk strategy. The categories were determined by the percentage allocated to cash (low risk), bonds (moderate risk), and equities (high risk). This allowed us to look at pairwise correlations between the fund scale variable and the variables of gender, age, marital status, spousal pension, and salary. The results showed that there was a statistically significant correlation between the fund scale variable and gender and salary.

We then estimated a linear regression model with the Q28 fund scale variable as the explanatory variable and gender, age, marital status, spousal pension, and salary as the independent variables. The model showed that salary was the only significant predictor of risk propensity as expressed by the allocation choices represented in the fund scale variable (see Tables 8 and 9 in Clark and Strauss, 2008). We were intrigued to note that the explanatory variable of gender nonetheless had a weaker association with the fund allocation choice. This led us to explore the actual pattern of allocation choices made in Q28.

---

[14] There is considerable debate in the academic literature over the relevance or otherwise of the size of bet in gauging respondents' solutions to these types of risks. Here, the issue is framed in a conventional manner, following the relevant literature in psychology and economics. However, one of us has sought to show that there may be money-effect on decision-making utilizing relevant stakes set against social status (Clark et al., 2009).

We conclude that the stylized profiles provide sufficient evidence to suggest the importance of asset allocation by social groups. The subsample sizes were problematic, therefore rather than modeling the asset allocation strategies we decided to examine graphical representations of choices for select groups. In Figure 3.1, the asset allocations of 4×2 types of people are summarized so as to illustrate how interaction effects can give rise to rather different investment profiles by people's social identity. The information is provided in the form of box-plots of asset allocation representing the range and the quartiles of the data (amounts allocated to each asset class) and the outliers.[15]

Figures 3.1A and B illustrate differences in risk propensity as represented by the choice of assets among low-income individuals, married and single, with no spousal pension. There were notable differences between single and married low-income young men. The former display a preference for higher allocations to cash with a median allocation in the range of 50 percent and the distribution skewed toward the right—that is, higher allocation values—with greater variability and lower allocations to bonds and equities, though in all cases the maximum allocation value is 100 percent and the minimum is zero.

For married low-income men, the preference was for bonds with a median allocation of 40 percent and a right-skewed distribution. It is striking that there is much less variability but more outliers in this group. Most young, married, low-income men have similar allocation patterns, but a few display preferences that are markedly different. There is much more variability among young unmarried men of the same socio-economic group, but fewer instances of extreme difference. Notice, moreover, that for married men the maximum allocation of 100 percent and minimum of zero only occur as outliers. Not only do members of this group display less variability in their asset allocation strategies but they also allocate more evenly, with typical allocations ranging between 20 and 60 percent in all three asset classes (compare with Benartzi and Thaler's 2002 1/n heuristic).

Married individuals with median incomes were the largest socio-demographic group within our sample. Here, elder, married, middle-income women display an aversion to equities (Figures 3.1C and D). The median allocation is 20 percent, but the lower quartile range drops to zero. The asset allocations between young, married, middle-income women and men are markedly similar, with women displaying a slightly stronger preference for cash and less variability overall but more outliers, and with men showing more values in the lower quartile range but more variability (up to a maximum of 100 percent) for cash and more variability (but fewer outliers) overall. Both

---

[15] The box contains the central 50 percent of the distribution of asset allocations ranging from the lower to the upper quartile. The line within each box represents the median and the whiskers indicate the maximum and minimum values. Where they occur, outliers are represented by single points.

Risk Propensities

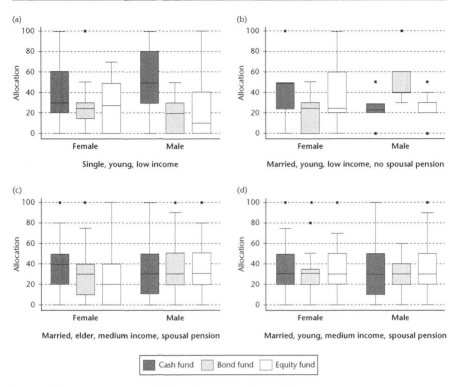

Figure 3.1. Asset-allocation choices for selected socio-demographic groups

Note: The sample sizes in the four diagrams were: A (11 males, 25 females), B (8 males, 12 females), C (95 males, 64 females), and D (61 males, 46 females)

groups had a median allocation of 30 percent for equities, but with maximum allocations of 70 percent for women and 90 percent for men and outliers of 100 percent for men and women. In both cases, the range dropped to zero for minimum allocations. Interestingly, these results suggest that while women are more risk averse than men (noted above), young, married, working women with pension entitlements choose asset allocation strategies close to those of similar men. These asset allocation strategies by socio-demographic status merit further investigation with a larger sample, such that statistical tests could be used to validate the implied scope of our findings.

Implications and Conclusions

This chapter provides a snapshot of the risk propensities of UK occupational pension participants. It highlights the significance of respondents' socio-demographic characteristics for understanding individual risk propensities

55

and putative asset allocation. Based on a random survey of the British population, it was shown that respondents' risk propensities did not vary in a statistically significant manner according to either the means of retirement saving or the type of pension plan. This is an important finding in that DB and DC pension plans have very different characteristics as regards the distribution of risk. Many pension plan participants do not appear to appreciate their relative insulation from, or their vulnerability to, the risks of different pension plan types. This finding suggests that UK pension plan participants are often ignorant of the basic structure of different types of retirement saving institutions.

Our results also suggested that respondents' socio-demographic characteristics are important in producing systematic patterns in pension-related risk propensities. First, our results confirm that women are generally more risk averse than men, while suggesting that older people are less risk averse than middle-aged people (compare Legros, 2006). Second, although the results do not suggest that marital status alone is a significant characteristic in discriminating between those with more or less risk tolerance, the existence of a spousal pension entitlement does have a statistically significant effect on risk tolerance. By this logic, when both partners have a pension entitlement they appear to pool risk between them—a household effect, no doubt, and one that mirrors findings from the United States (Friedberg and Webb, 2006). Third, this effect varies among married people of different ages and income levels, suggesting the existence of collective retirement planning by households able and willing to do so.

Fourth, these results point to a degree of social sorting by partner pension entitlement not often acknowledged in the pensions literature. These patterns have a distinct gender dimension, which may have significant implications for some women's long-term retirement welfare. It was shown that women with a pension entitlement are likely to have spouses with a pension entitlement, although the reverse only holds true for younger middle-income men. Furthermore, it is apparent that women who are in jobs that allow them to accrue pension benefits tend to be married to men who also have jobs with pension rights. This finding is in keeping with the literature on the increasing inequality *between* women, due in part to marriages between men and women with higher education levels and high earning potential (see Costa and Kahn, 2000 and Compton and Pollack, 2007 on "power couples").

We conjecture that women who end up with pension entitlements have selected partners who have also ended up with pension entitlements, thereby increasing long-term household income while pooling longevity and retirement income risks. Of course, this sorting process by spousal pension entitlement is likely to be part of more systematic processes of social sorting through partner selection and household formation; we do not mean to imply a level of individual foresight and utility maximization in the choice of partner that we would otherwise deem unlikely in financial planning and decision-making, that is, people don't just choose their partners for their pension potential! Nonetheless, the implications of sorting are clear. Low-income women face the greatest risk of poverty in retirement (see Bellamy and Rake, 2005; Ginn et al., 2001) and are the only group for whom a spousal pension entitlement does not produce a negative marginal probability of risk aversion. This kind of sorting has the potential to exacerbate the growing income gap *between* women in regard to their partners' short- and long-term potential incomes (see generally McDowell, 2004).

If these results were carried over to the entire UK population, there may be three implications for the design and implementation of supplementary pensions. The first follows from the fact that our respondents do not appear to distinguish their risk propensities by the means of retirement savings and the type of pension plan: on this most basic of issues, plan sponsors may need to improve their education and information programs. If important for those already enrolled in a pension of a certain type, it is crucial that plan participants be able to distinguish between plan types when confronted with the choice of continuing with an employer plan or opting out, and with a DB pension as opposed to the offer of a DC alternative. Given that DB pension plans are closing and employers are seeking to redistribute the risks associated with offering pension benefits, confusion on these two counts could significantly disadvantage many people. This highlights the essential need for information and education programs that recognize the danger of such confusion and help the public reach informed decisions (see, more generally, Cutler, 1997).

A second implication is that the design of education and information programs may benefit from a greater appreciation of the socio-demographic characteristics of those involved. Many programs treat plan participants as *individual* decision-makers with idiosyncratic preferences and interests; many programs are not attuned to the role of gender, age, income, and household status in affecting risk-related attitudes and behavior. Most importantly, to the extent that these characteristics systematically influence the risk propensities of particular types of people, there may be a role for designing education and information programs to suit their audiences.

The third implication of our analysis is that current pension education and information programs may be less well geared to those who plan for the future within the context of the household (spousal pension entitlements). Similarly, education and information programs that treat participants *as if* socio-demographic and spousal pension entitlements were irrelevant run the risk of glossing over the plight of those with the biggest stake (low-income women) in planning for an adequate retirement income while being inappropriate to the more sophisticated household retirement income strategies of higher-income men and women.

Finally, our results have relevance for the design of the National Pension Saving Scheme (NPSS) or the National Employment Savings Trust (NEST). The Pensions Commission recommended the establishment of the scheme, recognizing the decline in private sector occupation pensions combined with low savings rates especially among low-income workers. Even if not compulsory, the Commission suggested that an automatic enrolment scheme that relied upon the double-default predilections of many people—an unwillingness to exercise their opt-out options and an unwillingness to exercise their active investment options—could significantly increase private retirement savings and improve equitability. There are many aspects to the plan to be determined (not least of which are how and when it would be implemented). However, given our respondents' lack of knowledge about the risks associated with competing sources of retirement income and types of pension plans, there are reasons to support Thaler and Sunstein's call for "libertarian paternalism" (2003).[16]

At the same time, NEST will allow for plan participants to exercise their risk preferences through investment. As in DC schemes, NEST would presumably rely upon the individual for decision-making (ignoring household structure), would be neutral with regard to socio-demographic characteristics (including gender and age), and would provide a low-risk default investment option for those unable or unwilling to make risk-related decisions (matching the high risk aversion of low-income women). Just as private DC schemes may need to calibrate their education and information programs to the socio-demographic status of participants, so too NEST may have to take greater responsibility for the prospective welfare of those groups most reliant on NEST for supplementary savings. By this assessment, "libertarian paternalism" may be a recipe for accentuating long-term income inequality (see our detailed assessment in Chapter 8).

---

[16] See Cronqvist and Thaler (2004: 424–5) who describe this notion as follows: "(t)he idea is for the program designer to create an environment in which unsophisticated participants are gently guided in a manner that is intended to make them better off without restricting the freedom of the more sophisticated participants."

4

Sophistication, Salience, and Scale

Western governments have sought to encourage working men and women to take greater responsibility for retirement planning. For many writers, this responsibility reflects larger forces at work including the long-term transformation of twentieth-century welfare states (Esping-Andersen, 1999), the increasing importance of markets for the provision of social goods (Smith and Easterlow, 2005), and the financialization of everyday life (Boyer, 2000; Langley, 2008). From a continental European perspective, this "new" responsibility could fracture social solidarity and unravel stakeholder capitalism (De Deken et al., 2006). From an Anglo-American perspective, greater individual responsibility is more likely part and parcel of long-term structural changes in economy, culture, and society (Preda, 2004), epitomized by the decline of defined benefit pension plan coverage rates and the growth in private sector defined contribution and personal pension plans (Munnell, 2006; Strauss, 2008a).

There is a "premium" on planning for retirement to be reaped by those willing and able to plan, whatever the state of financial markets. For some theorists, people are ready and able to plan for the future—their apparent self-interest in realizing life-time consumption goals is believed sufficient to motivate planning now and in the future.[1] In the bedrock of standard microeconomic theory, some years ago Deaton (1992) raised doubts about the extent to which observed patterns of income, consumption, and saving were consistent with the permanent income hypothesis (compare Scholz et al., 2006). Similarly, those writing at the interface between psychology and economics have raised doubts about individuals' decision-making competence in the context of risk and uncertainty (Kahneman and Tversky, 1979). To the extent that people have well-defined

---

[1] This is, of course, yet another instance of the strongest form of the rational actor model. As we suggested previously, the virtue of the behavioral revolution has been its questioning of competence rather than disputing whether people are, or are not, rational in a substantive sense. See also Doherty (2003).

discount functions, many still find it difficult to marshal the requisite attention and cognitive resources to carry through on plans for the future (Ainslie, 2005).

Inspired by Lusardi and Mitchell (2007, 2008), in this chapter we tackle a subset of issues at the heart of this research. Based upon the previously described 2006 random sample of UK working men and women aged 18–59 years who participate in employer-sponsored supplementary pension schemes, we estimate the correlates of (*a*) the importance attributed by respondents to retirement planning, (*b*) the degree to which respondents believe they are prepared to plan, and (*c*) their knowledge of annuities (a financial instrument that converts accumulated savings into a guaranteed income stream until death). In doing so, we clarify what is meant by pension planning with implications for the structure of our subsequent analysis; we explain the logic of the statistical procedures used to produce the results; and we offer an interpretation of the results, emphasizing the sophistication of financial judgment, the salience of the issue for respondents' welfare, and the geographical scale at which respondents make planning decisions.

Matching the results reported in Chapter 3, we find that the older the respondent, the higher their income, and the degree to which they recognize pensions are designed to supplement retirement income, the more likely they are to believe planning is important, to be prepared for planning, and to be knowledgeable about annuities. The scale of effective retirement planning is the household; respondents with a spouse claiming similar entitlements are more likely to believe planning for retirement is important. By contrast, remote sources of planning advice and information carry less weight for our respondents than expert advice relevant to the circumstances of respondents' households. These findings reinforce arguments by Strauss (2008*b*) about the importance of social differentiation *and* scale for behavior under risk and uncertainty. These results also give weight to those who argue for a "relational" theory of economic life albeit embedded in personal relationships of commitment or contract (see Bathelt and Glückler, 2005).

If profound questions have been raised by the global financial crisis about the financial integrity of private funded pensions, this chapter provides, nonetheless, a benchmark against which we can judge the behavioral responses to these cataclysmic events now and in the future. In Chapter 5 we report on the investment intentions of defined contribution pension plan participants who, by virtue of their market location, were able to anticipate the consequences of the bubble (and bust) in UK housing prices.

Rationality and Planning

While there is little direct evidence that social status is correlated with certain types of behavioral anomalies and biases, there is evidence that consistency of decision-making is correlated with domain-specific skills and expertise (Wagner, 2002). This finding has implications for the quality of individual financial decision-making, including the selection of those responsible for long-term investment (Clark et al., 2007). Unclear, at present, is whether people are cognitively predisposed to perform more or less well certain types of tasks and functions (Kahneman's 2003 argument), whether common sense and experience can make up for cognitive shortfalls (Gigerenzer's 2004 argument for the significance of heuristics), and whether education can be effective in self-regulating acknowledged behavioral biases (Doherty's 2003 optimistic scenario). Nonetheless, we can hypothesize that, all things being equal, a resource-rich environment can compensate for people's acknowledged shortcomings in behavior and/or knowledge and experience (as suggested by Hershey et al., 2008).

In cognitive science, planning for the future is similarly assumed to be a human trait. In fact, it is widely believed that this cognitive function is located in the frontal lobes of the brain wherein self-control and ancillary functions provide subconscious mechanisms for governing individual behavior. Often invoked at this juncture are case-studies, and in particular the celebrated case of one Phineas Gage who experienced frontal lobe trauma, survived, but then exhibited quite erratic behavior (Libet, 2004).[2] For some, this trait has its origins in human evolution—specifically the claimed innate commitment in favor of the conservation of human life (a selfish interest) and/or the protection of progeny (perhaps an altruistic interest). We note, as well, the premium placed on time-dependent behavior traced by historians and geographers back to the industrial revolution and to contemporary market imperatives driving the measurement of time and planning for the future. Calibrating the value of time is a practice deeply embedded in everyday life (Glennie and Thrift, 2005).

Whatever the contemporary significance or otherwise of planning, we can *generally* agree with Bratman (1987: 8) when he says that "we are planning creatures. We form future-directed intentions as parts of larger plans, plans which play characteristic roles in coordination and ongoing practical reasoning; plans which allow us to extend the influence of present deliberation to the future." However, we must take care to be clear about the scope of our agreement and where there may remain disagreement about the significance of intention and the scope of planning.

In much of the social science literature, planning is seen as a means to an end that combines deliberation with an over-arching future-oriented

---

[2] A detailed description of the case and the implications of Gage's head trauma for understanding the status of foresight and planning are to be found in Damasio (1994) and Damasio et al. (1994). The interaction between canonical cases and the tried and tested evidence gleaned from controlled experiments is important in psychology and cognitive science, if not the social sciences such as economics, politics, and (to an extent) economic geography.

optimizing framework. We ought to be cautious about the generality of this framework for three reasons. First, just because we are "planning creatures" does not necessarily mean that we are conscious of related actions; planning could well be a finely-honed intuition which is used day in and day out without agents ever being aware that this is what we *should* do in situations X, Y, and Z. Second, planning need not be an intuition or a fixed repertoire of actions that originates with the person who takes action; planning could involve the imitation and emulation of others' actions and outcomes, such that the individual concerned neither deliberates over actions to imitate nor consciously makes a choice of those whom he or she wishes to emulate (Hurley, 2008). Third, planning may be an illusion; it presupposes control over the relevant decision variables and related environment (Wenger, 2002).

If planning is simply a process of matching day-to-day imperatives with a repertoire of possible actions *or* a largely unproblematic process of imitating others' actions given their place in our immediate field of vision, there would seem to be much less to planning than imagined by philosophers and social scientists. But if planning is responsive to the salience of the issue, and demands the sophisticated application of knowledge outside commonplace routines, then planning may well claim significant but scarce cognitive resources such as attention and deliberation (see Gabaix et al., 2006). As such, we should recognize that people are not equally endowed with either the cognitive resources or social resources to make good on intended actions and outcomes.

The Planning Process

We can classify human planning by reference to the nature and significance of the issue (salience), distinguishing between the routine and the important, and the nature and significance of the cognitive resources used in the planning process (sophistication), distinguishing between the use of "fast and frugal" intuition techniques and deliberative techniques. So, for example, in Figure 4.1 the upper left-hand box identifies a type of planning (Type 1) that is both commonplace and largely incremental in its positive and negative effects. By contrast, the bottom right-hand box identifies a type of planning (Type 4) that is demanding in terms of its use of cognitive resources and possibly very important in terms of its long-term effects on human welfare. There are, no doubt, people who occupy the spaces set aside for Type 2 and Type 3 planning, deploying scarce cognitive resources to make routine plans (being cautious) and responding to significant issues without the benefit of conscious deliberation (being impulsive).

		Sophistication of Judgment	
		Intuition	Deliberation
Salience of Issue	Routine	Type 1 (incremental)	Type 2 (cautious)
	Important	Type 3 (impulsive)	Type 4 (long-term)

Figure 4.1. Classification of planning types with reference to the salience of issues and the sophistication of judgment
Source: Authors (noting the inspiration provided by Bratman, 1987 and Wenger, 2002).

But what drives the planning process? Standard treatments of the topic begin with beliefs and desires—being states of mind that have, presumably, three characteristics (Bratman, 1987: 22): they control conduct, are relatively consistent or stable over the flow of life-events, and provide the impetus for action. There is no need for beliefs and desires to be anything more than a set of prior commitments which may or may not be recognized as such by agents. For example, a basic belief/desire may be self-preservation. Another belief/desire may be the love and respect of others. From beliefs and desires come intentions to realize those interests. By this logic, intention is the transfer function between motivation and action, wherein the latter is the expression in concrete form of the former. For intention to be effective, people need the cognitive resources consistent with realizing their interests as well as the social resources to sustain the chosen action or actions that are intended to realize underlying beliefs and desires. There are many instances where people's beliefs and desires are frustrated by poor reasoning and a lack of resources.

Bratman's model of "common-sense" planning is amenable to the application of optimization methods, and it has a recipe for public policy. But whether it is, in fact, consistent with the evidence as to how people make and execute plans is open to dispute. One problem has to do with the origin of beliefs and desires. By the previous account, beliefs and desires are distinct from cognitive processes, being primal urges, emotional commitments, or metaphysical incantations. For some, beliefs and desires reside in the mind while cognition resides in the brain. But, as we know, this is a categorical mistake (Ryle, 1949). In any event, recent research in the cognitive sciences disputes the plausibility of a sequential order to the planning process as set out above. Libet (2004) contends, on the basis of experimental evidence, that we act before we recognize that we have decided to act. "Intention" may be backward looking (as justification for action) rather than forward looking (as the point of initiation of action).[3]

This is not to say that consciousness and deliberation are superficial behavioral traits more related to self-justification than calculated action. In fact, the experimental evidence suggests that consciousness and deliberation play vital roles in governing behavior, especially in cases where agents take the time to check on their initial impulses or plans of action and revise according to their circumstances (Bratman, 2007). In this sense, Type 2 and Type 4 planning are second-order versions of Type 1 and Type 3 planning, being sophisticated expressions of self-conscious deliberation where salience (at some threshold value) governs the level of attention paid to the issue at hand (see Pacherie, 2006). Whether people articulate the differences between these types of planning is debatable. Those who hold to a strong version of rationality believe that people have a ready grasp of their relevance and act accordingly. Others are not so sanguine, as indicated by the logic of the *Interim Report* of the UK Pensions Commission (2004a).

Many people neither appreciate the importance of pension saving nor are equipped to make informed decisions, especially when dealing with complex financial products such as annuities (Poterba, 2006). Idealizing planning capacity may be a recipe for naïve decision-making (Type 1), self-defeating decision-making (Type 2), and short-term decision-making (Type 3). Employers, concerned about their responsibility for the decisions made by employees, may auto-enroll employees and offer opt-out options to those self-confident about their knowledge and understanding of the issues (i.e., Type 4 planners). As we know, most people do not exercise their opt-out right, just as most participants in employer-sponsored pension saving schemes do not exercise their right to make investment decisions. There may be Type 2 planners who claim the right to deliberate and decide; recognizing this possibility, the employer could make the opt-out option difficult to understand and

---

[3] See Bratman (2009), where he acknowledges that his normative model of planning may fall foul of the empirical evidence concerning cognition and behavior. We are reluctant to take sides on the issue, recognizing that many people come to issues such as planning for the future with a repertoire of decision-aids that short-cut deliberation in favor of response (see Chapters 1 and 2). There are those who dispute both the integrity of the experimental evidence and the implications for consciousness drawn thereof (see Mele, 2009). We are drawn to the evidence rather than categorical statements regarding the profound significance of consciousness for human beings.

time-consuming to execute, such that impulsiveness is constrained by the structure of the decision-making process.

In effect, those who use auto-enrollment to structure employee pension-saving decision-making take advantage of the fact that most people are on automatic pilot when it comes to planning for the future. Likewise, those who advocate stepwise time-dependent escalation devices to increase the rate of pension saving appear to assume that most people are Type 1 planners. That is, it is assumed that most people can accommodate incremental changes in pension contribution rates within the mix of issues they regularly respond to on a month-to-month or year-to-year basis. Benartzi and Thaler's formula (2005) for employer-sponsored defined contribution schemes assumes that most people are NOT Type 4 planners (being one justification for employers and government to take a more active and paternal interest in the long-term consequences of people's choices).

Framing the Empirical Analysis

The application of conscious deliberation to retirement planning depends, in part, upon agents' recognition of the significance of the issues at hand. In this chapter, we consider whether our survey respondents appreciate the significance of retirement planning, have the necessary resources to plan, and have the knowledge and understanding appropriate to purchasing annuities—a type of financial product directly relevant to preserving the accumulated value of retirement savings through a guaranteed income stream (Sheshinski, 2008). Here, as noted in Chapter 3, we use a random survey of relevant respondents to test their views on these three matters, evaluating opinion against their socioeconomic status and risk-preferences—proxies for the resources available for making planning decisions.

Previously it was shown that there are systematic age and gender effects on risk attitudes and putative asset allocation decisions (see also Clark and Strauss, 2008). Similarly, we found that gender, income, and household status interact such that, in relation to these decisions, men are more risk tolerant than women, older higher-income men are more risk tolerant than middle-aged higher-income men, and younger lower-income women are more risk adverse than older higher-income women. These findings are similar to those found in other countries such as Germany (see Dohmen et al., 2005). We also found, moreover, a significant "household effect" in that men and women who have a spouse with some form of pension entitlement engage in risk-sharing within the household. It would appear that they "plan" either in response to their circumstances (weighted by gender, age, and income) or by deliberate policies of risk management with respect to desired long-term household retirement income (also weighted by gender, age, and income). Here, we interrogate this implication directly with tests of both the importance attributed to planning and respondents' preparedness.

About eighty questions were asked of respondents. After an introductory question which sought their agreement to the interview (Yes or No), those agreeing to participate were asked two "qualifying" questions. S1: *"First, can I just check, are you currently working in an organization with at least 10 employees worldwide?"* Yes—Continue; No—Terminate. S3A[S]: *"The survey concerns pensions and benefits at work; do you currently have any form of pension?"* Yes—Continue; No—Terminate. Thereafter a series of questions were asked to establish respondents' type or types of pension plans, excluding those who indicated that they currently participated in defined benefit or final salary schemes. In doing so, it was intended that the survey concentrate on those for whom their own pension planning would have tangible benefits: those who by the nature of the employer-sponsored pension plan have personal responsibility for risk-taking and the expected value of the accumulated pool of retirement assets.

A set of questions established respondents' gender, age, location, job type, household status, annual income, and spousal pension entitlement. Question 9[S] asked respondents to strongly agree, agree, neither agree nor disagree, disagree, strongly disagree, or N/A with the statement: *"Planning for my retirement is an important issue for me."* Thereafter a series of similar statements were posed designed to elicit respondents' preparedness or otherwise for long-term retirement planning. For example, statement 10[S] was *"I know where to get more information about planning for my retirement if I need to"* and statement 12 [S] was *"my understanding of the savings and investment options available for long-term financial planning is reasonably good (or better)."* Based on a string of eight related statements, an index of preparedness was constructed using the response options noted above to create a respondent-specific score (explained in the Appendix).

A set of questions were posed to elicit respondents' attitudes to risk, investment, and annuities. As noted previously, risk tolerance was determined by asking respondents (Question 25[S]) to choose either *"I aim to get the best possible growth in the value of my savings, even if that means taking some risks which could cause my savings to fall in value"* (designated risk tolerant) or *"I prefer to have safe and secure savings and investments, even if that means they do not grow in value as much as they could"* (designated risk averse) or N/A. Mixed in with a series of questions regarding their knowledge and understanding of different types of investment funds including cash, bonds, and equities was question 27[S]: *"If you retired today, you would be required to use your pension savings to buy an annuity. How good is your understanding of what an annuity is?"* Respondents were given the following response options: very good, good, neither good nor poor, poor, very poor, and N/A.

In this chapter, we also sought to determine if there was a discernible *scale* (individual, household, and regional) at which pension planning takes place (capitalizing on Sunley's 2000 regional map of UK pension potentials).

Summary of Findings

We began with a model designed to predict respondents' belief in the importance of planning, using the ordinal logistic regression estimation technique. The results are explained in detail in Clark et al. (2009), including reference to the estimated parameters. Where possible in this chapter we keep the discussion of findings relatively general, leaving the details to the published paper (available from the authors).

As widely recognized in the literature, we found that AGE was statistically significant, with parameter signs and estimated values indicating that the older the respondent the more likely they believed planning for the future to be important. It was found that GENDER was statistically significant, though at a weaker level than other significant variables, suggesting that women were less likely than men to believe pension planning to be important. INCOME was also statistically significant, with the likelihood of indicating planning to be important strengthening with income. By contrast, it was found that neither REGION of residence nor the expected REPLACEMENT rate of income supplements due to pension saving were statistically significant.[4] Below, it is shown that respondents' REGION of residence does appear significant in other cases but not to the extent found in our analysis of annuities (see Chapter 5).

As indicated above, we sought to determine respondents' RISK tolerance in saving, using a question that gave a dichotomous choice. It was found that RISK tolerant respondents were more likely to believe that pension planning is important. We also posed a related question about personal responsibility for asset allocation. Question 48[S] began by noting that it is "sensible" for people to shift their assets out of "high risk assets such as equities" as they approach retirement. Respondents were given five response options including: (*a*) leaving the choice and strategy with the respondent, (*b*) leaving the choice with the respondent but with provider advice on strategy, (*c*) leaving the decision with the provider, (*d*) don't know, and (*e*) N/A. This is a significant issue, having implications for the conservation of accumulated wealth and the

---

[4] Region was coded using respondents' STD identifiers, based upon the Office of National Statistics UK regions. See http://www.statistics.gov.uk/geography/downloads/GB_GOR98_A4.pdf

translation of pension savings into an annuity.[5] Respondent attitudes with respect to the desired level of responsibility for age-related retirement-saving management were NOT significant for either the importance of pension planning, or preparedness, or their knowledge and understanding of annuities, and this is dealt with at length in Chapter 7.

As in Chapter 3, it was found that respondents with a SPOUSE also entitled to supplementary pensions were more likely to believe that pension planning is important, just as those respondents who indicated that they will RELY upon another for retirement welfare were found to be less likely to believe that pension planning is important. As for information in pension planning, those who used INFORMATION sources other than a pension specialist were less likely to believe pension planning is important. This was especially apparent for those who used more geographically remote sources of information, such as the Internet and booklets. Interestingly, it seems that bank advisors were less effective in this respect than insurance company representatives.

To understand the implications of our findings better, we considered whether the variables could make a significant difference to the results in interaction with one another—using probabilities. For example, on the SPOUSAL variable, the probability of being in the strongly agreed category that planning is important increased by 10.2 percent when respondents had a spouse with a pension entitlement. The predicted probability of being in the strongly agreed category that planning is important increased by 22.2 percent if the respondent's AGE was 50 years or more. For ordinal variables, the predicted probability of being in a response category was calculated using a 1 unit increase against the mean of the ordinal predictor. The probability of strongly agreeing that planning is important decreased by 11 percent with a 1 unit increase in RELYING upon another (see Table 4.1).

Turning to planning preparedness, AGE and INCOME were significant (as before), although it was found that the parameters on younger age groups were not significant, while it was found that the parameters on lower-income groups were significant. As anticipated, the older the respondent and the higher their income the more likely they would indicate that they were prepared for pension planning. In this instance, however, other variables were significant, including the parameter on REGION (but just the southwest), and, importantly, respondents' expectations as to the income REPLACEMENT rate of their pension savings. That is, the higher the expected

---

[5] Answers to this question have significant implications for the design of defined contribution savings schemes that use asset allocation rebalancing systems dependent upon participants' age and expected years to retirement (Bodie and Treussard, 2007). Whether participants take an interest in the "glide path" to retirement and whether plan sponsors should institute systems that do so automatically are issues to be resolved in the design of such systems (especially in the United Kingdom; see Byrne et al., 2007).

Table 4.1. Probabilities of significant variables in the ordinal logistic regression of planning importance

Significant variable	Planning is important				
	Strongly disagree	Disagree (%)	Neither agree nor disagree (%)	Agree (%)	Strongly agree (%)
Socio-demographic variables					
Age 30–39**	−1.14	−7.3	−1.5	+10.2	
Age 40–49***	−2.0	−11.4	−6.0	+19.5	
Age 50 or older***	−2.2	−12.6	−7.4	+22.2	
Female*	+0.4	+2.5	+2.9	−5.9	
Having a spouse with a pension***	−0.6	−3.8	−4.6	+9.1	
Income £40,001– £65,000*	−0.5	−3.6	−9.2	+13.4	
Risk/information variables					
Pensions Info: insurance company**	+0.7	+4.3	+8.0	−13.0	
Pensions Info: a booklet***	+1.2	+7.6	+5.9	−14.8	
Pensions Info: the Internet*	+1.0	+6.6	+9.7	−17.3	
Pensions Info: bank advisor***	+0.7	+5.0	+9.4	−15.2	
Savings risk**	−0.6	−4.1	−6.6	11.3	
Relying on another (1–5)***	+0.7	+4.7	+5.6	−11.0	

Strength of significance: *** = $p < 0.01$; ** = $p < 0.05$; * = $p < 0.1$.

value of respondents' pension savings against earned income, the more likely they would indicate preparedness for pension planning. In this case, GENDER was not significant.

Again, RISK tolerance was found to be highly significant and positive in effect, although the parameter value was less than half that found in the previous analysis. And once again, respondents with a SPOUSE entitled to a pension benefit were positively disposed to planning preparedness while those RELIANT on others for retirement income were more likely to indicate less preparedness. Consistent with the previous findings, compared to a pension specialist those reliant on other sources of INFORMATION were less likely to be prepared for planning. In this case, though, many more sources of information were found significant and it was difficult to discern a pattern in terms of the estimated value of the relevant parameters. Intriguingly, those who would rely upon a colleague or relative and those who would rely on a manager in their organization for information were less likely to indicate preparedness than those reliant upon booklets and the media. In the next section, we suggest one interpretation of this finding: a colleague or relative might be their first "port of call" after respondents' spouses.

Finally, in anticipation of our study of annuity patterns (Chapter 6), we sought knowledge of the correlates of respondents' knowledge of annuities. These were consistent with previous results in that AGE, INCOME, REPLACEMENT, and REGION (weak) of residence had the same signs as before and were significant predictors of the likelihood that respondents have knowledge of the structure and design of annuities. Also significant were RISK tolerance, INFORMATION sources, and RELIANCE on others for pension income upon retirement. But notice that neither GENDER nor SPOUSAL entitlement is significant, and INFORMATION sources other than a pension specialist were of limited number and significance. In this case, the two sources significant were a colleague or relative and a manager in the organization. Generic and institutional sources of annuity information were not statistically significant.

To give a better indication of the strength of the coefficients, we also calculated predicted probabilities of significant variables. For the binary variable RISK, the estimates predict the probability of having a specific category of knowledge of an annuity when the binary variable is active (holding all other variables constant at their mean). For RISK, the probability of having a good knowledge of an annuity increases by 15.4 percent when respondents indicated RISK tolerance. For nominal variables, the predicted probability is again specified with an "if" statement for each category. For example, the predicted probability of indicating a "good" knowledge of annuities increases by 19.7 percent *if* the respondent's AGE is 50 years or older. For ordinal variables, the predicted probability of being in each category is calculated with a 1 unit increase (centered around the mean) in the ordinal predictor. So, the probability of having a good knowledge of annuities decreases by 2.9 percent with a 1 unit (mean-centered) increase in RELYING on another.

Sophistication, Salience, and Scale

Comparing results across the three issues considered, Table 4.2 provides a list of the independent variables, their statistical significance by issue and the sign on the estimated parameters. There was a high degree of consistency in results: the significance of selected independent variables and the signs on estimated parameters. Further, there were some interesting results. For example, while the literature on financial decision-making suggests that men and women have different levels of risk tolerance (risk aversion), the GENDER variable was only weakly significant for the importance of planning and insignificant otherwise (see Bajtelsmit, 2006). Whereas we show in Chapter 5 that in the intended take-up of annuities in the aftermath of the 1990s bubble and bust the REGION of residence was a significant factor in discriminating between

Sophistication, Salience, and Scale

Table 4.2. Statistically significant correlates of pension planning; parameter signs, strength of significance, and consistency across the indicators of planning capacity

Independent variable	Pension planning*		
	Importance	Preparedness	Annuity
AGE**	+ve	+ve	+ve
GENDER	−ve (w)	x	x
INCOME**	+ve	+ve	+ve
REPLACEMENT rate**	x	+ve	+ve
REGION of residence**	x	+ve (w)	+ve (w)
RISK tolerance	+ve	+ve	+ve
SPOUSAL entitlement	+ve	+ve	x
RELIANCE on others	−ve	−ve	−ve
INFORMATION source**	−ve	−ve	−ve

* Where (w) refers to statistical significance at the 0.10 level and x indicates no statistical significance;
** indicates that the variable is set against a base-case.

respondents, here the effect was weaker and less systematic. In all cases, the signs on the estimated parameters were the same and most estimated parameters were significant at the 95 or 99 percent confidence levels.

We now turn to interpreting the results. Like others writing on financial planning and pension saving (see Agarwal et al., 2008), our interpretations are suggestive rather than definitive. Further, as argued above, care must be taken not to over-interpret the results in the sense of attributing intentionality to each and every factor in the pension planning process. In doing so, we distinguish three types of effects or aspects of pension planning, beginning with RISK tolerance. It has been noted many times that people are generally risk adverse—a core empirical finding underpinning the behavioral revolution in the social sciences (Baron, 2008). Here, it was found that those indicating RISK tolerance as opposed to risk aversion were more likely to believe pension planning to be important, were prepared for pension planning, and had good or very good knowledge of annuities. This suggests that financial *sophistication* may be a precondition for deliberate pension planning.

From Table 4.2, it is also apparent that AGE, INCOME, and recognition of the role of pension savings in the REPLACEMENT of earned income were significant. The older a respondent the more likely he or she believed pension planning to be very important. If intuitively plausible, this result does not accord entirely with theoretical expectations derived from the permanent income hypothesis (see Deaton, 1992). It seems that as people come closer to retirement they are either more conscious of the need to plan for the near future, or they are more likely to recognize the costs of not having adequately planned for retirement in the past. The INCOME effect reinforces the AGE result: either people seek to conserve their accumulated well-being, and/or they fear for their standard of living over the near future.[6] The effect of REPLACEMENT on retirement planning is simpler: given the similar size of statistically significant coefficients of different replacement rates, it would seem that it is sufficient to recognize that pension saving is designed to replace earned income. Each variable is indicative of the *salience* of pension planning (see also Gabaix et al., 2006).

Perhaps the most interesting set of significant predictors of pension planning are those that refer to the *scale* of decision-making. As we noted previously, some respondents are clearly influenced by the existence of a SPOUSAL pension entitlement. This could mean that there is mutual learning between partners.[7] As such, the relevant retirement income planning unit could, in fact, be the household not the individual. It could also mean that individuals are aware or become aware that their long-term welfare is a function of their partner's survival prospects (see generally Friedberg and Webb, 2006).[8] Reinforcing this result is the fact that some respondents intend to RELY upon others for their retirement, discounting the value they attribute to their own pension planning while implicitly or explicitly placing a premium on others' best intentions *and* planning capacity.

That there is such an intimate *scale* to pension planning is a significant result compared to the relative lack of significance attributed to REGION of residence. Just as important in this respect, however, are the results on the sources of pension planning INFORMATION. Against the base-case, a pension specialist, other sources of pension planning information were found to be less likely to contribute to a sense that planning is important and that respondents were prepared for planning. Notably, an insurance company advisor and bank advisor were deemed to be more likely to prompt respondents to suggest that pension planning is important than geographically remote sources of information such as the media and the Internet. It was harder to draw this implication from the results on preparedness, although it should be

---

[6] It is possible that older, higher-income people have a better sense of what can be achieved by pension saving because of their life-time success and/or the fact that there is a clear cause-and-effect relationship between pension saving (now) and retirement income (near future) given a declining capacity to earn higher income. See Wenger (2002: ch. 3) on the psychological relationship between individual experience and the perception of agency which is amplified by the consistency of success.

[7] An intimate interest in another's pension prospects may prompt discussion of the implications of different pension saving options and the costs and benefits of alternative courses of action. Each may think it necessary to justify his or her plans, thereby linking in a deliberate fashion plans with intended outcomes (separately and together). See Pettit (2007) on the virtues of discursive deliberation.

[8] This result suggests the existence of implicit or explicit tontine-like household pension planning. A tontine is a club for mutual insurance wherein those who begin the club bet on their survival against other members such that as club members die those surviving receive larger and larger shares of income until the person who outlives all others gets the benefit of all the members' initial investment. It was a common form of "pension" saving in England during the Middle Ages and has some relevance even today. See Sheshinski (2008) and Goldsticker (2007).

noted that a pension specialist is clearly more important than a relative or colleague.

On annuities, it would seem that the only other source of information of value to respondents was a colleague or relative, but note the small size of the parameter. We would contend that the significance attributed to pension specialists is a similar effect to SPOUSAL entitlement and RELIANCE on others—in this instance, though, regulated by contract and fee-for-service, rather than intimacy.

We would also contend that there is evidence for another consistent but negative scale effect. According to our respondents, generic information sources such as the media and the Internet are not as valued as sources of information that come from direct contact with advisors, albeit less knowledgeable about the specific issues related to pension planning. This finding suggests that a "national money guidance service" as recommended by the Thoresen Review (2008) may be ineffective unless remote sources of information are complemented with personal advice from truly independent professional sources.

Implications and Conclusions

Too often, biological predisposition is presumed to extend through to actual behavior, ignoring the cognitive and social resources necessary for effective decision-making as well as the role that context plays in channeling behavior (see Goldstein et al., 2001). In any event, we have suggested that too much is made of conscious deliberation, invoking recent research in the cognitive sciences to suggest that "planning" can be the expression of routine signal-response mechanisms that govern human behavior; people may plan, but do so without the conscious deliberation often assumed by theorists.

There is a type of planning that is a self-conscious check on impulse and intuition—even if Kahneman (2003) believes impulse and intuition to be the best available solutions to decision-related tasks, pension and retirement saving decisions carry important implications for long-term welfare. As a consequence, it is crucial for public and private institutions to set prompts and incentives for people to take account of the likely results of their planning decisions (however arrived at) (see Benartzi and Thaler, 2005). More optimistic analysts, writing in the wake of the behavioral revolution, argue or seek to show that most people develop appropriate and informed if not optimal solutions to common planning problems (Doherty, 2003; Gigerenzer, 2004). In this case, it may be useful to know about the correlates of respondents' attitudes to pension planning as a guide to designing effective public policy.

In this chapter, we showed that there are three types of significant correlates of pension planning. There are correlates, such as risk tolerance, that are indicative of respondents' financial *sophistication*; as in many other studies of expertise and intelligence, those who understand the substantive foundations of financial decision-making may be, all things being equal, more likely to recognize the importance of pension planning (compare with Wagner, 2002). Just as important are factors that drive the *salience* of pension planning for respondents, matching a related argument made by Gabaix et al. (2006) to the effect that salience is crucial given the cognitive and resource costs of deliberation. Those who recognize that pension planning is important are those who have the most to lose from not planning. Perhaps this reflects Kahneman and Tversky's finding (1979) about the importance of loss aversion in decision-making under risk and uncertainty.

As for pension planning, *scale* counts: that is, with household relationships and with knowledgeable advisors but not at the workplace (compare Bernheim, 1998). There is a certain intimacy to these particular relationships that adds value to respondents' confidence in pension planning. By contrast, respondents tended to discount the value of generic sources of information compared to specialist advisors. This is, of course, consistent with the fact that skills and expertise are domain-specific; by training and education or by virtue of the tacit knowledge embedded in certain roles and responsibilities, it is not surprising that respondents value pension specialists over bank advisors and insurance company salespersons. Whether respondents do, in fact, recognize this logic is impossible to determine. They may distrust advisors with an obvious interest in selling pension products and services by the companies that employ them (ignoring the fact that most "independent" pension advisors are actually on retainer to large financial service companies).

In terms of public policy, three implications deserve mention. Following on from the discussion immediately above, there are reasons to doubt the utility of generic financial advisory services (compare DWP, 2006). If discounted by those respondents most likely to believe that pension planning is important, these services are, at best, ignored by those without the sophistication needed to make informed judgments about the nature and quality of proffered advice. At worst, generic advisory services may encourage unwarranted confidence in those who can least afford to make errors in pension planning. A second and related implication is that those people able and willing to make plans for the future may benefit from government policy designed to improve the quality and quantity of advice offered by specialist services. This may mean requiring disclosure by advisors about their relationships with financial service companies. A third implication is that the delivery of such advisory services may be best focused on the household rather than the individual or place of employment. At the margin, there is also evidence for considering regionally targeted advisory programs, even if we share with recent writers concerns about the effectiveness of financial literacy programs in general (see Atkinson, 2008).

APPENDIX

Variables in the Scaled "Preparedness" Index

To construct the preparedness for retirement planning index, a set of eight variables were developed from the survey questions and integrated into a single variable. Factor analysis was used to determine significant factors among the survey's knowledge, confidence, and ability to afford expertise variables. The following eight variables were loaded as a single factor with an Eigenvalue of 2.46. The eight variables were scaled into the index with a reliability test Cronbach's alpha of 0.78 and an average interim covariance of 0.29. Note that all variables used a five-point Likert response scale.

Variable name	Survey question
Planning Informed	I know where to get more information about planning for my retirement if I need to:
Confident11	I am confident that I am doing enough to make financial preparations for my retirement:
Knowledge15	I would like to do more to plan for my retirement, but I don't know what I should do:
Afford16	I would like to do more to plan for my retirement, but cannot afford to do so:
Confknowledge33	I feel comfortable that I have sufficient knowledge to choose how to spread my savings between these funds. [Referring to question 28: Imagine that you now have to invest your long-term savings, for the majority of years between now and your retirement. Please indicate how you would invest your money in the three types of funds listed, cash, bond, equity.]
Knowledge34	Choosing from this list of funds is too difficult and I need a simpler option:
Confknowledge43	In answering question 40, I felt comfortable that I had sufficient knowledge to make the choice between the funds available. [Referring to question 40: "Having read the new definitions above, again imagine that you have to invest your long-term savings for the majority of years between now and your retirement. Assuming these funds are the only available options and you must put all your money in one fund, which fund would you choose?"]
Knowledgeannuity	If you retired today, you would be required to use your pension savings to buy an annuity. How good is your understanding of what an annuity is?

5

Being in the Market

The house-price bubble came to an end in 2007 with the onset of the subprime credit crisis which morphed into the global financial crisis. Over the previous decade, UK owners saw their homes double or triple in price. Debate raged over the extent to which the boom was an asset-price bubble like the tech-bubble of the late 1990s. For those convinced that the housing boom was a bubble subject to pathologies of overconfidence and irrational exuberance, the bubble had to burst (Shiller, 2008). There remains, however, a long-standing shortfall in supply that marks the UK housing market as somewhat different from that of the United States. In the United Kingdom, the limited supply of housing, the demand for single- and multifamily rental accommodation, and the nature of urban regeneration in major cities reflected structural imperfections in property markets as well as a regulatory regime that truncated market response to burgeoning demand, especially in southern England (Hamnett, 2009).[1]

Recognizing the supply constraints, the "optimistic" long-term scenario for UK housing markets, emphasizing market stability and supply constraints underwritten by rising real incomes, held sway over pessimistic assessments (see Nickell, 2005; PricewaterhouseCoopers, 2006). Even Shiller (2007: 6–7) was circumspect about the "possible reversal in coming years" of the house-price boom, suggesting that, for the United States at least, "it may be hard to understand from past experience what to expect next, since the magnitude of the boom (was) unprecedented." As is now recognized, the "boom" was a "bubble" and the accelerating global financial disaster was actually closer to hand than many realized—witness the September 2007 crisis of confidence in the UK building society turned bank Northern Rock (National Audit Office, 2009).

---

[1] Affordable housing in relation to real incomes was a crucial issue for the UK government over the last phase of the house-price boom. See Hamnett and Whitelegg (2007) and the HM Government (2007) "green paper" on supply and the prospects for affordable environmental sustainability.

It could be argued that a financial crisis was believed so unlikely that all market participants were caught up in the emotion driving UK and global housing markets (Christie et al., 2008). In this chapter, however, it is demonstrated that the housing bubble could be seen in the pattern of UK housing prices compared to the paths of real income growth and stock market movements over the previous two decades. It is contended that those best able to assess market patterns and make reasoned judgments distinguishing the bubble from a boom relied upon the tacit knowledge available to skilled and experienced financial market participants. Gertler (2003: 78) notes that tacit knowledge is produced through practice and "the relationship between tacit knowledge and context is a reflexive one, since tacit knowledge both defines, and is defined by, social context." Our proposition is that being in the London financial market *and* having the knowledge and experience to judge the relative risks of property with respect to other retirement investment instruments were crucial ingredients in managing the risks of planning for the future.

We begin by mapping the nature and scope of market behavior, building upon the insights of the behavioral revolution with respect to time–space myopia. In doing so, we extend our taxonomy of planning to provide a schematic model of market participants, emphasizing the role that knowledge, experience, and skill may have in generating different investment strategies among those similarly located. To provide an empirical base, we focus upon the London employees of a large multinational investment bank who either directly or by default sought to diversify the risks associated with property in their retirement savings and investment portfolios. Whereas commentators debated whether the United Kingdom was in the grip of a speculative bubble through to the onset of the global financial crisis or not, it would appear that a number of our respondents sought a "safety first" strategy before the public at large realized the high costs associated with property as a pension-saving option. While our respondents may not have anticipated the depth and consequences of the looming crisis, their intended investment strategies were consistent with a sophisticated approach to retirement planning.

Empirically, we consider three related issues: first, the degree to which components of UK working income were correlated with the path of house prices and stock market prices over the past twenty years; second, the degree to which house prices could be said to have exhibited bubble-like characteristics; and third, the degree to which the employees of a London-based international investment bank expressed a preference for property in their retirement portfolios. Given the claims made about the value of property as a desirable form of retirement investment leading up to the crash, it could be hypothesized that respondents had strong preferences for property in their investment portfolios. Testing for patterns of response, it was found that few respondents would have relied, in fact, upon property. Respondents' age, income, and risk tolerance were significant factors in discriminating between similarly placed individuals in terms of their commitment to property as a form of retirement investment. Those respondents who would have property at the core of their pension investment portfolios seemed to do so either because of limited options or because their lack of experience let the gamble run on.

Our respondents were deeply embedded in the world of global finance wherein the vast majority indicated a "good" or "very good" understanding of property and managed investment funds (compared to the financial knowledge and understanding of "average" UK employees; see Chapter 4). Important for this chapter is the empirical demonstration of bubble-like characteristics in the pattern of house prices, providing a crucial reference point for understanding the sophistication or otherwise of our respondents. This finding is useful, as well, in making informed argument against those who held that the housing market was caught up in more profound problems, such as the unsustainable lending practices of major financial institutions. Coming at the peak of the property bubble and before widespread appreciation of the enormous costs of the global financial crisis, our research shows that there were sophisticated investors attuned to market risks across the range of retirement investment instruments, even though they may not have applied that judgment to their bank's retail and wholesale investment funds.[2]

Taxonomy of Market Behavior

As previously noted, the research program on the efficient markets hypothesis was premised upon the assumption of rational agents and, hence, risk-pricing. This assumption allowed analysts to ignore "less than rational" behavior, based on the argument that either other market agents take advantage of those not able to recognize and respond to market signals effectively, or market agents are not self-defeating. Either way, it was assumed that arbitrage drives markets to efficiency (Stein, 2009). As a consequence, there is little in the efficient markets hypothesis that can help to explain the systematic coexistence of different types of more or less skilled market agents. If tacit

---

[2] For reasons of confidentiality we are not able to identify the name or national origin of the investment bank. Readers will be interested to learn that the company was not, initially, as badly affected by the global credit crisis as some other institutions, and certainly did not face the predicament of Lehman Bros and other banks that were forced into mergers with other institutions or were underwritten by their "home" governments in the form of large equity stakes. Like many Western banks, though, its current situation is problematic.

knowledge were significant for market agents, it is axiomatic that this information would drive the formation of market prices; any premium on tacit knowledge would be discounted over the long term.

Rational expectations are difficult to "prove," and attempts to estimate market prices based upon its arguments require ad hoc adjustments so as to approximate observed patterns (Arrow, 1986). With the "governing" principle of finance, there have been attempts to introduce more nuanced conceptions of market behavior, including the application of Simon's (1956) notion that people, more often than not, satisfice rather than maximize, and that people find it difficult to process information on a timely basis because of innate cognitive limits. As evidence accumulated demonstrating the existence of cognitive biases and anomalies, it also became apparent that financial markets exhibit systematic patterns of performance at odds with theoretical expectations (Shleifer, 2000). The behavioral revolution gained traction, in part, because of the availability of large databases and the perceived commercial advantages of anticipating market behavior (see Shiller, 2000).

The core empirical finding of the behavioral revolution led by Kahneman and Tversky (1979) among others was that people are doubly myopic: people tend to be short-term oriented *and* they tend to rely upon the immediate environment for clues about responding to unanticipated events. This finding has been produced many times, often relying upon university students as respondents and the shared protocols governing experiments and testing procedures (Baron, 2008). How and why this is the case is contentious. In Chapter 4, it was noted that analysts often invoke customary habits of mind and the costs of conscious deliberation (Gabaix et al., 2006; Kahneman, 2003). Comparing university students with financial decision-makers who have significant roles and responsibilities, we have also found strong evidence of myopia (Clark et al., 2006). But we also found with skilled and experienced respondents (as opposed to neophytes) that sophisticated decision-making techniques combined with context-specific knowledge can make a difference to solving financial problems (Clark et al., 2007).

It can be shown that there is a range of behavior in financial markets such that tacit knowledge when combined with the requisite skills can give some market agents leverage not available to other market agents. This does not mean that some agents are rational whereas others are not, nor does it mean necessarily that the coexistence of different types of market agents is not sustainable over the long term. For Kahneman and Tversky, rationality is a human trait; the debate over whether humans are rational or not, which has bedeviled the social sciences for many years, is not at issue. Rather, domains that are subject to risk and uncertainty demand a level of sophistication that goes beyond basic levels of competence, learning by doing, and learning by

		Time Horizon	
		Short-term	Long-term
Spatial Scale	Local	Myopic	Naïve planner
	Global	Opportunist	Sophisticate

Figure 5.1. Behavioral predispositions of market agents at the intersection between time and space with respect to financial decision-making under risk and uncertainty

interacting (Gertler, 2003). The environment or context of behavior matters a great deal for the effective performance of human rationality.[3]

To illustrate this argument, Figure 5.1 provides a schematic of the possible scope of human behavior in the context of risk and uncertainty. It is anchored by reference to "myopia" in the upper left-hand box, wherein people are transfixed by recent events that happen in their immediate activity space. To be myopic is to be parochial: to lack foresight and conceptual insight about the temporal pattern of market activity and the interaction between local events and market forces operating within and without the immediate environment. If tacit knowledge is at hand, myopic market participants are not able to make meaning out of that information. By contrast, sophisticated market participants are deemed to have the knowledge and experience to make sense of the flow of events, recognizing the interplay between what happens in local and global financial markets. Sophisticates have a long-term

---

[3] The value of "being there" is a combination of access to hitherto difficult to obtain information, superior judgment that comes with experience, and finely-honed heuristics that structure decision-making. It may also involve relationships with others that contribute knowledge and understanding of relevant issues. Having such a privileged position in relation to market dynamics may make for better long-term performance (against chance and the prospect of being right about market momentum in one period but not in other periods). Such a position may prompt, nonetheless, overconfidence in one's judgment—a common-enough behavior (see generally Baron, 2008 and more specifically Døskeland and Hvide, 2011). Likewise, overconfidence may be accompanied by taking a gamble, a possibility given related research on the predilection to gamble among educated and financially skilled individuals (see Clark et al., 2009). "Being there" is not a recipe for some kind of "super-rationality."

perspective and a conceptual understanding of market structure and performance (bottom right-hand box).[4]

According to the logic underpinning Figure 5.1, there may be investors who hold to a long-term perspective but do so without understanding the interaction effects between local markets and global markets. For Mullainathan (2007: 91), naïve planners tend either to hold too long to past commitments or abandon those commitments when events turn against expectations (upper right-hand box). Under- and overreaction to market movements is characteristic of many market players, just as a lack of perspective about the origin and transmission of market movements may mean that they respond to "local" signals without understanding the broader context. As important are the "opportunists" who focus upon short-term market movements, moving between markets as arbitrage opportunities are presented. These investors care little about tacit knowledge; their skill is all about being able to exploit minor differences between markets against expectations utilizing leverage in relation to markets' momentum (bottom left-hand box).

Hilton (2003) and Shiller (2000) contend that a predisposition in favor of recent events often translates into herd behavior, amplifying short-term market movements while threatening those who stay too long in the market with lower returns than those able to judge short-term movements against long-term local and global patterns. Nonetheless, myopia and opportunism are commonplace in financial markets made so, in part, by the incentives offered to the employees of financial institutions. This point is developed below. We should also observe, though, that to the extent to which market players are able to distinguish between the risks and rewards due to their employment and the risks and rewards attending their long-term welfare, market participants may be "opportunists" and "sophisticates" in different settings. See Clark et al. (2006), showing that pension fund trustees' risk "appetite" varies according to their responsibilities and their personal interests.

Finance, Contingent Income, and Risk

In a similar vein, Langley (2008) sought to bring global finance to ground: to the everyday networks of financial relationships and commitments that dominate individuals' lives. His point was that the geography of finance is simultaneously local and global, penetrating to the smallest nook and cranny of the

---

[4] Mullainathan (2007: 91) observed that sophisticates recognize how a long-term commitment to saving and investment must be governed in relation to short-term temptation and the possibility that plans may have to be remade to deal with unanticipated events that require recasting objectives.

Anglo-American world while being intimately linked with market arbitrageurs operating around the world. His examples were both mundane and compelling, linking, for example, individual deposit and savings accounts to worldwide processes of risk-pricing. In this domain, individuals not only must plan for the future, but they must also "price" the risks of competing financial institutions and instruments on a global basis so as to hedge the risks that they are neither able nor willing to carry into the future. In effect, their risk appetite and their capacity to manage those risks are constitutive elements of their long-term prospects.

For Langley, exposure to the market permeates everyday life to such an extent that people have come to internalize the norms of personal responsibility for retirement welfare. For some writers, this is representative of larger political forces that have conspired to discount the postwar welfare state and rewrite the social contract, putting in play "social goods" that were deemed to be best provided outside the market (see Smith, 2005). This resonates with continental European commentaries and criticism, being an issue that is part of a continuing debate over the reach of Anglo-American market institutions into the fabric of European society (Boyer, 2000; Engelen, 2003). In the Anglo-American world, markets play a greater role in allocating social resources than twenty-five or even ten years ago. Nation-states are dwarfed by the financial institutions and markets to which they play host, notwithstanding the pivotal role of government in bailing out failing financial institutions.

Were the UK housing boom just an issue of demand and supply, resolved in the manner of the housing boom of the late 1980s, this chapter would have been about market structure and performance. But over the past ten years or so, property has figured prominently in the public imagination as a valuable asset class for retirement saving and investment. As Strauss (2009a) notes, in the years leading up to the subprime crisis, the popular press was awash with stories to the effect that property is an attractive alternative to conventional employer-sponsored retirement schemes.[5] If highly problematic, given recent events in global financial and mortgage markets, the dramatic shift in the UK private sector from defined benefit (DB) to defined contribution (DC) and personal pension plans gave weight to the argument that individual exposure to the risks associated with securities markets ought to be balanced by other kinds of investments including property.[6] At issue, in part, is the extent to

---

[5] See Chapter 6 opening, on the *Sunday Times* interview with a "public" person as regards their spending, saving, and pension behavior. Commonly, they respond by arguing that property is better than a pension because the property market is easier to understand.

[6] It is important to recognize at this juncture that in DC plans, plan participants bear the risks associated with investing for their future retirement income. By contrast, in DB plans, the plan sponsor bears the risks associated with realizing the promised pension benefit (Clark and Monk, 2008). This means, more often than not, that DC plan participants are fully exposed to financial market performance—a reality that is quite profound for those enrolled in such plans and about to retire in the midst of the global financial crisis.

which respondents' retirement investment portfolios were diversified across a variety of more or less cross-correlated assets.

In this study, our respondents were DC plan participants. Few would have carried a DB plan into the bank, and few would have wished to be in a DB plan given anticipated employment prospects. In any event, their compensation packages combined a base salary with the promise of performance-related annual bonuses and stock options which were dependent, in part, on the nature of the jobs they performed, the company's profitability, and market movements. These types of compensation packages were very important in the finance and technology sectors for "incentivizing" valued employees (Teece, 2000), even though senior executives claimed large unfunded DB pensions upon termination. At the limit, our respondents were either directly involved in tournaments for jobs or sought to take advantage of the existence of these tournaments to enhance their bonuses. Not surprisingly, younger workers and employers believe DB pensions to be irrelevant to their short-term (but different) goals (see Clark and Monk, 2008 for empirical research on this point and Roberts, 2004 for a theoretical exposition).

Contingent compensation has also been identified as important in other UK sectors and in cohorts of workers typically younger than the baby-boom generation. Dickens (2000) showed that the younger a male British worker, the more likely he was dependent upon a "transitory" element for his earned income (compared to the "permanent" component attributable to human capital). Dickens also showed that this transitory component persists over time, being important in explaining yearly increasing levels of earned-income inequality in British society. Although his characterization of occupation was crude, he demonstrated that individuals from so-called white-collar professions exhibited a larger transitory component compared to individuals from the so-called blue-collar occupations. Extended by Kalwij and Alessie (2003), research suggests that there is a market premium for younger white-collar workers due to happenstance or social position, and that this premium accentuates measured income inequalities as the effect of human capital dampens with age.[7]

The culture of contingent compensation has been about immediate reward and long-term leverage in favor of personal wealth. It is, or was, a male culture of risk-taking personally and professionally that has had as its goal the joint maximization of short-term returns (subject to the parameters set by the firm

---

[7] This is one side of the polarization of UK real wage growth over the past twenty years. See Goos and Manning (2007) for the definitive treatment of UK wage inequality and their framework applied to the London labor market by May et al. (2007). Immigration has also played an important role in the recruitment of labor to the London financial industry at both ends of the wage and skill spectrum (see McDowell et al., 2008, 2009).

in relation to its market position).[8] Few employees expected to be with the same firm in five years' time, and even fewer expected to be in the industry at age 50. As it happens, there were significant long-tail (negative) risks in these situations. One was unknowingly accepting counter-party trades that exploited the bank's risk controls thereby endangering job tenure and bank solvency. Another was being seduced by the prospect of wealth such that base salaries were willingly discounted in favor of "promised" windfalls. Yet another was the possibility that salaries *and* bonuses were so highly correlated with market movements that retirement investment portfolios had the same risk profiles as current and expected earned income.

Patterns of Earned Income and Market Volatility

Bonuses in the finance industry were paid against market performance, implying an autoregressive (AR) year-to-year carryover effect, as indicated by run-ups in stock market performance. There were also significant stochastic elements (MA), as illustrated by unexpected market movements over the quarters since the 2007 peak of the boom. Dickens (2000) and Kalwij and Alessie (2003) suggested that the persistence of the transitory component has tended to increase over time, especially for younger age cohorts. Looking back over the past twenty years, and in particular the period 2001–7 which saw the run-up in UK house prices, in this section we report on whether (*a*) growth in earned income and market movements can be segmented into distinct periods such as 1988–2000 and 2001–7; (*b*) whether their paths can be approximated by statistically similar ARIMA (autoregressive integrated moving average) processes; and (*c*) whether growth in income, housing, and stock market movements over the period 1988–2007 can be shown to have been significantly cross-correlated.

Stability of mean annual growth rates

A set of income series was collected from the Office of National Statistics (ONS) on individuals' quarterly earned income, other earned income including returns on investment and property, and total income (sum of the two). As for securities markets, the daily FTSE100 index was used to represent the path of stock price movements and the monthly Nationwide Building Society House Price index was used to represent the path of housing prices. All series

---

[8] McDowell (2008) provides a revealing assessment of gendered roles and responsibilities. Note, though, that our results do not hinge on gender so much as age and income (where, of course, most older higher-income employees were men).

were transformed into quarterly rates of growth or log differences and discounted by the Bank of England consumer price index.[9]

As explained in Clark et al. (2010), the mean annual growth rates for all five series were displayed, with a distinction made between the whole period 1989 (Q1) to 2007 (Q2) and the two subperiods pre-2001 and post-2001. A *t*-test was used to determine whether to reject the hypothesis that series' mean real growth rates were statistically the same before and after 2001. We found that this hypothesis could not be rejected for three of the five series, despite the fact that income growth rates were higher in the post-2001 period and volatility in each of these series was lower in the post-2001 period, whereas the FTSE100 growth rate was much lower (negative) with a higher variance.[10] The hypothesis of similarity was rejected for the FTSE100 and house prices; there was a significant difference between the pre- and post-2001 growth in house prices, with a much lower variance in post-2001 growth rates. This relatively simple test of means and variance suggests a UK housing *bubble* possibly amplified by the stock market with implications for retirement portfolios.

Modeling income and market movements

The next step in the analysis was to estimate ARIMA models of the five series. As we should expect from Dickens (2000), it was found that the income series combined significant autoregressive terms with effects spread over three quarters in two of the three series and a significant and positive stochastic component two quarters prior to the present value. In this respect, earned income and other income had "memories" that tended to reinforce the path of growth in income as well as event-specific shocks that also contribute to growth. By contrast, the FTSE100 series contained only AR elements and specifically lags over one and two quarters.

UK house-price growth can be also approximated by an ARIMA model with significant AR terms distributed over three past quarters and a past quarter positive shock parameter. Given indications of bubble-like growth rates subsequent to 2001 in the difference of means test, we also estimated ARIMA models for house prices before and after 2001. It was found that the post-2001 series was not stationary and could be modeled as a random walk. A test of the boundary sensitivity of these results, shifting the first episode to 1988–99 and redefining the second episode as 2000–7, did not change the

---

[9] Data was found at http://www.statistics.gov.uk, http://www.uk.finance.yahoo.com, http://www.nationwide.co.uk/hpi, and http://www.bankofengland.co.uk

[10] Not reported, but available from the authors, are the results of a *t*-test on series' standard deviations which showed that the variances of four of the five series were found statistically different between subperiods except for the FTSE100.

Table 5.1. Correlation matrices of macro-variables, quarter-to-quarter: 1988–2007

	Total income	Wage income	Other sources of income	UK house index	FTSE100
Whole sample: 1988–2007					
Total income	1.000	0.857*	0.931*	0.487*	0.184
Wage income	0.857*	1.000	0.610*	0.586*	0.158
Other sources of income	0.931*	0.610*	1.000	0.335*	0.171
UK house index	0.487*	0.586*	0.335*	1.000	−0.126
FTSE100	0.184	0.158	0.171	−0.126	1.000
Pre-2001: 1988–2000					
Total income	1.000	0.874*	0.934*	0.657*	0.228
Wage income	0.874*	1.000	0.641*	0.772*	0.337**
Other sources of income	0.934*	0.641*	1.000	0.469*	0.109
UK house index	0.657*	0.772*	0.469*	1.000	0.515*
FTSE100	0.228	0.337**	0.109	0.515*	1.000
Post-2001: 2001–7					
Total income	1.000	0.716*	0.930*	−0.331***	0.341***
Wage income	0.716*	1.000	0.410**	−0.154	−0.060
Other sources of income	0.930*	0.410**	1.000	−0.348***	0.480**
UK house index	−0.331***	−0.154	−0.348***	1.000	−0.676*
FTSE100	0.341***	−0.060	0.480**	−0.676*	1.000

Significance level: *: 1%, **: 5%, ***: 10%.

results. We have evidence of a bubble in house-price growth over the period 2001–7, reinforcing the findings of Leamer (2007) and Taylor (2009) about the consequences for housing markets of highly expansionary monetary policy in the aftermath of the TMT (technology, media, and telecommunications) bubble and 9/11.

To determine whether the series were statistically related, we created a correlation matrix of all series based upon quarter-to-quarter growth rates. In Table 5.1, the correlations between series are displayed for the entire period, for 1989–2000, and for 2001–7, with their associated levels of significance. Clearly, the income series were highly correlated, with a higher variance in other income. More interesting was the fact that income series, house prices, and the FTSE100 were also statistically and positively cross-correlated for the entire period and for 1988–2000. These patterns are not so obvious in the second period 2001–7 when house-price growth was negatively correlated with income series and the FTSE100. The reader is reminded that all series were expressed in real terms, suggesting a lower rate of growth in real incomes compared to the high rate of growth in house prices. Significantly, over this same period other income was shown to be statistically and positively correlated with the FTSE100 index.

These results provide a number of insights and implications. First, there is clear statistical evidence in favor of a UK house-price bubble, 2001–7. Second, the bubble was expressed in higher rates of growth in house prices than real incomes (which were also growing but not at the same rate). Third, the coincidence and cross-correlation of series' growth rates prior to 2001 suggest an intimate relationship between earned incomes and financial markets, less evident though over the period 2001–7. Those sophisticated long-term planners, conscious of the whole-period relationship between series, could have seen positive cross-correlations between series as a prompt to diversify retirement investment. The negative cross-correlation between house prices and income in the second period might also have encouraged further investment in housing by those seduced by expectations—the housing bubble being amplified by its strong performance post-2001 relative to the performance of stock prices over the same period.

Retirement Investment and Portfolio Diversification

Notwithstanding the behavioral logic underpinning this book, the cross-correlations between income, stock markets, and property markets could be viewed through the lens of the permanent income hypothesis (Samuelson, 1969). Here, three threads of theory overlap to underpin this claim as a reference point for empirical analysis. At one level, the permanent income hypothesis can be invoked to set the context for decision-making both with respect to current and expected long-term consumption. At another level, a human capital model of earned income can be invoked to provide cohort-specific respondents a time horizon over which to estimate expected earned income. Further, a life-cycle model can be invoked to establish a mechanism for judging the significance of current events against expected income and plans for retirement consumption. In these ways, current events can be set within an overarching, time-dependent, long-term optimization framework with implications for current and future retirement portfolio design (Merton, 1969).

To make this argument work, however, heroic assumptions have to be made about the time horizon of behavior, individuals' discount rates, stable incomes, and market structure. Many question the plausibility of each and every element in this chain of reasoning. For example, Deaton (1992) raised doubts about the plausibility of the permanent income hypothesis; empirical evidence on individuals' discount rates do not support standard assumptions, prompting attempts to produce multiattribute models (Ainslie, 2001; Laibson, 2003); Dickens (2000) showed that current incomes have a significant stochastic component; and, more troubling, Heaton and Lucas (2000) note that standard models assume that income is not significantly correlated with stock returns and other assets such as housing, whereas we show above that there may be significant cross-correlations, especially in periods that carry high levels of risk and uncertainty. Even if people had the cognitive and social resources necessary to integrate time and space in a continuous optimizing framework, market structure would limit their ability to execute desired insurance strategies.[11]

Obviously, we are neither idealistic about behavior under risk and uncertainty nor are we convinced that people necessarily know and act according to their long-term best interests; see Chapter 4, showing that whether people plan for the future and the geographical scale at which they make plans are correlated with age, income, and household status. So what do people do in these circumstances? How do they utilize complementary mechanisms of investment in designing their retirement portfolios? How did they manage bubble-driven asset classes like property in relation to the other forms of retirement investment? Were most people myopic or were there sophisticated long-term planners who seemed to be aware of the risks of property investment in diversifying their retirement portfolios? These questions frame analysis reported in this section and in the following sections.

Our research is based on a specially designed and implemented May 2007 survey of employees from a large London-based multinational investment company at the peak of the bubble and before the onset of the global financial crisis. We drew younger, higher income, and higher educated respondents than is normally the case, all with responsibility for setting the parameters for long-term investment in their employer-sponsored DC pension plan. The questionnaire was designed to match, in part, our research on related issues of pension planning and decision-making in collaboration with Mercer Human Resource Consulting (explained in Chapter 3). Of approximately 7,000 London-based employees, 2,372 responded to the online survey. The survey was based on thirty-one structured questions that set common response options and sought information on conventional socio-demographic characteristics including income, age, gender, and household circumstances; expressed attitudes to statements of opinion or intention were normally assessed using a five-point Likert scale; and puzzles or problems to be solved focused upon asset allocation.

So as to establish a baseline against which to understand the significance of property for our respondents, the analysis proceeded in two stages focused upon the nature and scope of respondents' non-pension retirement savings. The following question was posed to all respondents: *"Apart from the company*

---

[11] Shiller's argument (1993) about the enormous welfare costs of incomplete markets applies in these circumstances. His argument was framed in relation to his experience in a housing boom and bust through the mid-1980s and early 1990s. His experience prompted a close examination of the psychology of markets among other topics.

pension scheme, what are the other ways by which you save for your retirement, if any? Please check all that apply." This question came early in the questionnaire following a series of questions brought together under the banner of retirement planning. Options for retirement savings included (in order of their significance for respondents): savings accounts (71.9 percent of respondents checked this option), other savings (such as personal equity plans (PEPs), individual savings accounts (ISAs), equities) (67.3 percent), occupational pensions from a previous employer (43.2 percent), property investments (e.g., buy-to-let, second home) (30.7 percent), other company-sponsored savings and/or deferred compensation schemes (20.0 percent), insurance-related savings (e.g., endowments) (15.2 percent), and other (9.0 percent).[12]

In the first instance, the dependent variable was the number of asset classes identified by respondents (0–7): that is, the diversity of their retirement savings portfolio, leaving aside their current employer-sponsored DC plan. Here, socio-demographic variables including age, income, gender, and household status were used as explanatory variables for portfolio diversification. Notice, though, that social scientists are properly cautious of inferences to the effect that socio-demographic status is causal: that is, by age, gender, and income people behave in predictable ways. Nonetheless, we showed in the previous chapters that age, income, and family status are closely related to the sophistication of pension planning, especially for older employees who have much to gain or much to lose by becoming involved in pension planning. As well, these variables have proven important in explaining patterns of risk propensities in asset allocation decisions. Having the confidence and capacity for pension planning is clearly important, especially in diverse populations with different levels of knowledge and understanding of the links between pension saving and the nature and purpose of various financial instruments.

Following on from these findings, we estimated the effects of respondents' attitudes and opinions on the number of instruments in their investment portfolios. This was done in the light of widely recognized recommendations by finance academics and professionals to the effect that the diversification of long-term risk across asset classes is one important function of a retirement savings portfolio (Sharpe, 2007). From modern portfolio theory (Markowitz, 1952) and behavioralism (Benartzi and Thaler, 2001) also comes the

---

[12] In the survey, we sought to distinguish respondents' owner-occupied homes from investment in the property markets, including tax-preferred buy-to-let housing. A home is a complex idea, being a proxy for children's desired schooling, a status good, a consumption item, an insurance against catastrophic illness, and an intended bequest to children (Smith, 2008). If it is a pension investment, this suggests that people intend to go into either the sale of the family home during retirement or some form of equity release. However, it is widely appreciated that neither options are preferred by current UK retirees. See the following chapter on the many interpretations attributed to the "family home."

hypothesis that an expressed preference for risk aversion is properly correlated with a larger number of asset classes (see more generally Litterman et al., 2003).[13]

The results of our analysis are presented in Clark et al. (2010: table 5). It was found that the closer the respondents were to retirement, the larger the number of investment vehicles; respondents with a spouse with a similar pension entitlement were associated with a larger number of investment instruments; and the more junior a respondent the fewer the number of instruments. However, being risk-tolerant was associated with more savings instruments, a finding not entirely consistent with theoretical expectations to the effect that risk aversion would be associated with a larger set of savings instruments.[14] Those lacking confidence in retirement planning tended to have fewer savings instruments, whereas those more confident had more savings instruments. It was difficult to interpret the finding that those who agreed or disagreed with the statement that they rely upon "someone else to support them" in retirement implied more savings instruments.

These findings were consistent with our findings for a representative sample of UK residents regarding the correlates of pension planning and the demand for consultation in pension investment. As noted above, it is not surprising that older people with spouses also involved in pension planning have a broader range of retirement investment instruments than others not so "located" (using Sharpe's 2007 term). This may be because they could afford to do so, or because they realized the importance of managing risk given pending retirement. Equally, the issue is more salient to them than is perhaps the case for younger, lower-income employees without the advantage of a joint interest in future welfare.

Property and Portfolio Diversification

We sought to determine whether there was anything distinctive about those respondents who expressed a preference for property, recognizing that about 25 percent of respondents were so inclined. In Table 5.2, we summarize the

---

[13] Risk management is framed here as an issue of portfolio diversification. As Paul Langley (pers. com.) has noted, this is a partial view considering that some respondents may have been parties to derivative products designed to dampen the risks associated with holding property. While possible, we would note that access to such products on an individual basis is rare, and very expensive.

[14] Those deemed risk tolerant (38.3 percent of respondents) were those who checked the option "*I aim to get the best possible growth in the value of my savings, even if that means taking the chance that I could lose money.*" By contrast, those deemed risk averse were those who "*prefer to have safe and secure savings and investments, even if that means they do not grow in value as much as they could.*" As noted in Chapter 3, this is a variation on Kahneman and Tversky's formulation (1979) of risk aversion, although not formally tested in an experimental situation.

Table 5.2. Property as a retirement savings instrument (company pension scheme)

	Observations	As fraction of the sample (%)
Only property or property and any other instrument	624	26.3
G1 Property	65	2.7
G2 Property + 1 other asset	138	5.8
G3 Property + 2 other assets	187	7.9
G4 Property + 3 other assets	133	5.6
G5 Property + 4 other assets	68	2.9
G6 Property + 5 or more other assets	33	1.4
Any number of non-property instruments	1,412	59.5
No other instruments	336	14.2
Total	2,372	100.0

pattern of a property preference, noting that only 65 of 2,372 respondents would have relied exclusively on property as an additional investment instrument to their company-sponsored DC plan. The table also shows that property in combination with other savings instruments was not particularly popular, recording 624 responses out of 2,372. Even so, there were some respondents who would have loaded up on property by limiting the combination of property with other instruments; the most popular combinations were property +1, property +2, and property +3 savings and investment instruments.

Determinants of a bias toward property

Here, the dependent variable was set as a binary variable, including property (= 1) or not (= 0). The results of a probit regression are reported in Clark et al. (2010). Curiously, it was easier to identify the negative case (excluding property) as opposed to the positive case (including property). The negative sign on the estimated parameters suggests that those separated and those in relatively low-status jobs, as well as those who agreed with the statement that the house they live in "will provide the majority of retirement needs" and those who agreed with the statement that they would like to do more for retirement but could not afford to do so, tended *not* to include property in their portfolios. These are plausible responses. Equally plausible are the positive results on risk tolerance and those who disagreed with the statement that they would rely upon someone else for their retirement. Harder to understand is the significant and positive parameter on gender (being female = 1) (compare with Watson and McNaughton, 2007).

In summary terms, those who were risk-tolerant and self-reliant for future retirement income were predisposed to include property in their savings portfolios, whereas there were those by dint of family status, (lower) job status, and disposable income not so predisposed. Remarkable was the finding that property was not a significant indicator (one way or another) aggregated by age, income, spousal pension, or recognition that planning for retirement is important. The variables shown as crucial to UK pension planning were not significant for the choice of property as a retirement savings instrument.

Property in combination with other instruments

What is the role of property in combination with other savings instruments? Can we identify correlates for property and the number of associated savings instruments? In this instance, our objective was to estimate the probability of belonging to any of the five groups focusing upon the statistical significance and sign (positive and negative) of the socioeconomic determinants. With respect to socio-demographic variables, we found age and income to be significant: the two older age categories, 40–49 years and 50–59 years (wherein the latter group has the largest parameter), and the two higher-income categories, medium and high income (wherein the former had the larger parameter value). The sign on age was positive, implying a bias toward property with more savings instruments, whereas the sign on income was negative, implying a bias toward property and fewer savings instruments relative to the five or more instruments. Clearly, significant in this case were the affirmative results on planning for retirement, being confident about doing enough for retirement, and relying upon others for support during retirement. In each case, the positive sign on "agreement" indicated a bias toward holding property with more savings instruments. Finally, risk-taking was also found to indicate a higher likelihood of holding property with more savings instruments.

Synthesis and Interpretation

Just as the housing bubble was about to implode, our survey sought evidence about the nature and scope of respondents' intended retirement savings and investment strategies. We were especially interested in their holistic approach to the issue, recognizing that few studies of pension investment step outside the parameters of offered pension schemes to include other components of household wealth. Most importantly, we sought to determine whether our respondents were sophisticated investors balancing the risks in income, securities markets, and property with other instruments not so obviously cross-correlated with short-term financial movements. Given their position in the London and global financial industry, one hypothesis was that their intended retirement investment strategies would take account of these risks and be appropriately diversified. The alternative hypothesis was that they too were caught up in the irrational exuberance attending the housing bubble, attributing great significance to property investment in their whole portfolios.

To sustain the analysis, we began with the macro-movements of national income, the London stock market and the UK House Price index. Here, as expected, we found significant and consistent levels of cross-correlation with evidence that the house-price boom was, in fact, a bubble amplified by the stock market. Given that compensation in the London financial industry carried a large "performance" component with rewards such as stock options, there can be little doubt as to the risk-exposure of our respondents to market movements. Having asked respondents to choose among seven savings and investment options other than the company DC pension plan, an important finding was that respondents were rather cautious about other investments, rating bank savings accounts above PEPs, ISAs, and equities, with due regard given to the pension plans of previous employers (43.2 percent). Property came fourth in the list of options, being preferred over other company deferred savings schemes and insurance products. Only 65 of 2,372 respondents chose to concentrate on property, though 458 of the 624 who identified property as a desirable option would have held limited numbers of other investment instruments.

Assuming that a sophisticated investor carries a diverse portfolio of retirement investment instruments in circumstances where current income is highly contingent on market movements, we sought to identify the determinants of portfolio diversification. As expected, those who would carry a larger number of instruments were older, had a spouse similarly engaged in pension planning, were confident about pension planning, and were risk-tolerant. Those who would carry a limited number of instruments held rather junior positions, earned average incomes (referencing bank standards), and lacked confidence in pension planning. It is arguable that those with diverse portfolios were those who recognized the importance of pension planning by reason of their age and immediate prospects (regardless of income). Equally, it is arguable that those with a more limited set of instruments were unable to manage their long-term risks, notwithstanding their recognition of the apparent risks involved.

Of those respondents who chose property, the significant positive coefficients were gender, self-reliance, risk tolerance, and (lower) income. This suggests a group of people who were rather self-confident about their own ability and/or a group of people who believed that property was the only viable option (given their limited options). The negative coefficients were, once again, those in relatively junior positions and those who indicated that they were simply not able to afford to save for retirement. What of the coefficients on property-centered diverse portfolios of retirement instruments? Once again, those respondents who combined property with a larger number of instruments were older, recognized that planning for the future was important, were confident, and were risk-tolerant. It was more difficult to characterize those who combined property with a limited number of instruments.

It is significant that age, income, and planning capacity all figured in affecting the probability of being in any of the five groups. This result suggests a level of conscious deliberation and choice wherein property and its related instruments were chosen for strategic reasons having to do with forecast and pending retirement. Those disenchanted or unable to plan by virtue of a lack of confidence or a lack of capacity did not appear to use property in a strategic manner.

We would contend that many of our respondents were sophisticated investors. Even if they could not have known the full extent of the global crisis in mortgage markets and the resulting decline in UK and global housing prices, let alone the cataclysmic events of late summer 2008, respondents took a safety-first approach to long-term risk (see Roy's 1952 seminal paper). Those most at risk to market turmoil and those with most to gain and lose from market movements in relation to their immediate plans for retirement appeared to be the most savvy investors; older, confident retirement planners had the broadest portfolios and tended to combine property (to the extent they deemed it a valuable instrument) with other instruments.

By contrast, there were some respondents who by dint of a lack of alternatives seemed to concentrate on property. If they recognized the likely risks associated with the housing bubble, they may have had few other options given their age and income, stage in careers, positions in the firm, and lack of savings. "Safety first" is an option only if respondents have the resources, experience, and judgment to back their retirement plans. Without options, even those conscious of the looming costs of the "bubble" may have been trapped in the property market.

Implications and Conclusions

In this chapter, we focused on the retirement savings and investment strategies of well-placed individuals, being both employees of a large London-based multinational investment company and participants in that company's DC plan. Whereas many studies of portfolio diversification consider the range of options and choices made within the bounds of DC plans, we looked outside the employer-sponsored plan to the whole retirement portfolios of employees, noting that recent research has suggested that a complete picture is necessary when seeking to understand individual retirement savings. Emphasized, as well, were the cross-correlations between current income, the stock market, and the UK housing market, hypothesizing that sophisticated investors would dampen the concentration of risks by diversifying their retirement investment portfolios.

Modeling macro-movements in incomes and markets, we showed that the bubble was apparent in the path of house prices over the period 2001–7. Many of our respondents pursued a safety-first retirement savings strategy, thereby either deliberately or in response to market pressure discounting the transfer of market and income risks into their retirement investment strategies. Property was not a significant component of many respondents' strategies; very few of our respondents concentrated on property as a desired form of retirement savings. Our respondents were savvy investors, and were not caught up in the irrational exuberance that seemed to dominate expectations of the proper role of property in UK residents' retirement plans (see Smith, 2008; Strauss, 2009a). Against public expectations, we would suggest that their safety-first retirement savings strategy reflected tacit knowledge and experience in financial markets.

We noted that age, income, confidence in retirement planning as well as risk tolerance were significant determinants of a diverse savings portfolio, whether across all retirement savings instruments or with respect to a property-centered savings portfolio. By contrast, junior employees (by job classification), lower-income employees (against the standards of the finance industry), and those less confident in retirement planning and not advantaged by having a spouse similarly placed with respect to pension entitlement tended to have less diverse savings portfolios and, in those cases where property was important, more reliance on property to the exclusion of other instruments. It could be argued that these results amplify and are consistent with the argument in Chapter 4 to the effect that salience and sophistication tend to be associated with pension planning; those by dint of age and income who have most to lose from rash investment decisions are those most focused on pension planning and investment decision-making.

But there is an unresolved issue of cause and effect embedded in the results. Are older, higher-income individuals more effective retirement planners and risk managers because of their apparent reliance upon investment decisions taken now and in the immediate future, *or* are they more likely to pursue the safety-first option because they are older with higher individual and household incomes? Does age and income enable effective retirement investment and planning because they have the resources to do so (unlike younger, more junior employees who tend not to have the benefits of higher incomes and accumulated wealth and resources)? Many of our respondents were younger. Compared to their peers in the industry, their base salaries were lower, even if they would have benefited from bonus and stock-option windfalls. If better off than the average UK worker, disposable incomes were likely to be highly constrained by expenses associated with their children's education, mortgage payments on the family home, and the repayment of other loans. It could be that their savings portfolios were limited by their commitments in ways that 40- and 50-year-olds are not.

Saving for Retirement

Lacking significant savings, being unable to afford ISAs, PEPs, and related investments, and being nervous of the high cross-correlations between their earned income and the path of the stock market, they may have seen property other than the family home as the only alternative (especially when buy-to-let carries tax-preferred rental income). By this account, our respondents may have been savvy in two senses: at one level recognizing the ideal strategy *if* they had the savings and wealth to do so, and at another level knowingly taking a gamble on property, given it was the only option that might provide a pay-off large enough to switch gears later on in favor of the safety-first option. By this logic, they were the agents *in* social theory (see Preda, 2004), even if unlikely to be so, given the findings of our research on the age-bias of pension planning.

Consequently, we conclude with the four types of market players introduced in Figure 5.1. There were sophisticated investors who discounted the related risks by a safety-first investment strategy; opportunistic investors who took a gamble on the path of house prices, relying upon inside information to cut and run from the market as soon as it turned down; naïve planners caught up in the irrational exuberance of the moment, notwithstanding a range of intended long-term savings options; and myopic investors, who by lack of income and lack of knowledge of the cross-correlations between macro-movements in financial markets overloaded on property and stayed too long in the property market, following the market down as they had followed the market up. Among our respondents, all four types of investors appear to have been present. But it would seem that most were either sophisticates or opportunists. Few were naïve or myopic—a finding consistent with Gertler's thesis of "being there" (*contra* O'Donoghue and Rabin, 1999).

Ultimately, this chapter can be read as an empirical demonstration of the argument advanced in Strauss (2009*b*) to the effect that there is an intimate, even reinforcing, relationship between behavior and context (or what Simon (1956) and Sharpe (2007) have referred to as the "environment" in which behavior takes place). It demonstrates that "being in the market" can reap dividends. It also suggests that the location of market agents, whether amateurs or experts, can frame the conceptualization of risk and set the options available to those agents. Of course, there are larger forces at work, including the transformation of what counts as state and individual responsibility for planning for the future. In part, this is a political project set within structural constraints, including the aging of developed economies' populations. One implication from our research is that "being in the market" may reinforce emerging income inequalities in earned income by amplifying those inequalities through the differential responses of people to market movements. These issues have been barely acknowledged by governments in their rush to embrace the new order of personal pension planning.

6

Housing, Retirement Saving, and Risk Aversion

Each week on the back page of the Money section of *The Sunday Times* (London), a minor celebrity is interviewed about his or her financial circumstances. The questions posed typically include how much cash the celebrity has to hand, how much they earned over the past year, their investments including property, and their intentions regarding saving for retirement. More often than not, the celebrity is asked to choose between a pension and property (sometimes identified as the family home) when saving for retirement. In the lead-up to the peak of the financial bubble (May 2007), many of those interviewed indicated that property was preferable to a pension. When respondents explained their choice of property over a pension, they often indicated that property is "easier to understand" than a pension.[1]

There is no doubt that UK property has been an attractive investment over the past two decades. Economic growth, limited supply and constraints on the planning approvals process, increasing household incomes, and the fact that housing has become a crucial instrumental variable for gaining access to high-quality social goods like education have conspired to accelerate the long-term trajectory of urban property prices. As well, it is apparent that there was a bubble in property prices over the first decade of the twenty-first century in a number of Western economies including the United States (Reinhart and Rogoff, 2009: fig. 13.2), Spain, and Ireland (Conefrey and Gerald, 2009). In Chapter 5, we used time-series techniques to show that UK property prices "took off" after 2001 in a manner consistent with a financial bubble. We also

---

[1] Those interviewed in the Money section of *The Sunday Times* often confuse occupational pensions with personal pensions and, like many people, are probably also confused as to their differences (Chapter 3). We do not analyze individuals' attitudes vis-à-vis expected income from the British State Pension (BSP), although many current retirees rely on the public pension for a significant proportion of their income (Pensions Commission, 2004b).

showed that there was close correlation between the paths of property prices, earned incomes, and the prices of other financial assets like equities.

It is entirely plausible that people believe property is "easy to understand" because, if considered on a neighborhood basis, "bricks and mortar" carry location-specific risks. Moreover, "investor subjects" increasingly recognize the fungibility (or interchangeability) of housing wealth (see Langley, 2006; Smith, 2008; Strauss, 2008c), not least because of the promotion of equity release schemes and the prospect of paying for social care in old age (Terry and Gibson, 2010). We hypothesize that relying upon property, and in particular one's home, for future retirement income is consistent with risk aversion. However, it would appear that property investment and reliance on one's home for future retirement income has been a risky strategy if we take into account the systemic risks correlated with global financial markets. If people are myopic about the scale and scope of the embedded risks, risk aversion at the "local" level may be misleading about the true nature of risks faced by individuals and their families (Smith, 2009). The sophistication or otherwise of individual savers, both with respect to understanding the scope of the risks associated with different retirement assets and with respect to understanding these risks in the context of market volatility, are crucial issues. In any event, whether investing in property was, or is, more risky than relying upon an occupational or personal pension depends a great deal upon the nature of the pension: defined benefit (DB) versus defined contribution (DC).

In this chapter, we look more closely at the preferred role of respondents' homes for a group of participants in a DC employer-provided pension plan. Previously, we considered the preferred role of investment property distinguishing that from the family home given the characteristics of the latter in relation to the former. Here, we are particularly interested in whether respondents, in relying upon the family home, could be characterized as more or less risk averse than those who would not rely upon the family home. Our data come from the survey we discussed in Chapter 5, which was conducted at the peak of the bubble (May 2007) and before the onset of the global financial crisis. Because respondents came from an institution that was clearly "in the market," and were thus more knowledgeable of the costs of financial volatility, we assume that our respondents were more aware of the risks associated with housing than the average person. As such, our analysis provides insight into whether there is, or was, a trade-off between contributive pensions and the home, and whether those who choose the latter over the former can be characterized as more or less risk averse.

By exploring the link between pension saving and the home, we need to explain the significance or otherwise of this possible connection.[2] Readers

---

[2] See also Pensions Policy Institute (2009) for an empirical analysis of the possible value of housing in underwriting retirement in the United Kingdom. They suggest, as we do, that a

may not be aware of the nature and significance of DC pensions for UK private sector employees (especially the finance sector which dominates London). In the following section, we summarize the risks embodied in DC pensions and how those risks appear to be correlated with other risks faced by employees (plan participants). The global financial crisis exposed the overlapping risks associated with current income and future income. Thereafter, we look more closely at the role that property plays in retirement portfolios, distinguishing between property investment per se and the family home. This is followed by a brief account of the survey used to elicit employee preferences especially as regards their conception of risk. We present the results of statistical analysis and their interpretation before drawing implications for understanding pension saving.

Three Pillars of Retirement Saving

Western societies arrange the retirement saving of their citizens through three pillars (World Bank, 1994): pillar 1, social security and insurance; pillar 2, occupational or employer-sponsored pensions; and pillar 3, individual or personal pension saving, tax preferred or not. It is widely assumed that because the first pillar of social security is underwritten by the nation-state, it is the least risky form of saving for retirement, even if the promised value is typically some premium on basic needs and welfare (Clark, 2003). Nonetheless, there are significant national differences between the expected value of government-sponsored social insurance (compare the United Kingdom with the United States and much of continental Europe; see De Deken et al., 2006). Traditionally, underwritten by the employer, occupational pensions were conceived in order to top up social security, so as to bring retirement income close to the final earned income. If less secure than social security, it was assumed that supplementary pensions were more secure than individual saving for retirement—the risk of the former being borne by the employer, whereas the risk of the latter is presumably borne by the individual.

In combination, the three pillars of retirement saving were expected to provide successive generations with an adequate retirement income—a higher average standard of living than was the case prior to the Second World War. By the mid-1990s, the UK government believed that it had produced a winning formula for funding retirement income superior to that of continental Europe. However, the expected costs of demographic aging combined with global

"holistic" approach is needed to retirement planning, recognizing the diversity of assets that households may have at their disposal in funding retirement. They tend to emphasize equity release over other options.

economic turmoil have undercut the capacity of nation-states to deliver on promised benefits to future retirees. Most important for the United Kingdom, private sector occupational pensions have come under sustained attack, not only for the value of promised benefits but also for the assumption that employers can afford to underwrite promised benefits with automatic cost-of-living adjustment. Private sector occupational pension coverage rates have declined dramatically over the past fifteen years. Furthermore, employers have forsaken DB pension plans in favor of DC plans (to the extent that employers have replaced closed DB plans; Clark, 2006*a*).

DB pension plans promise a certain income upon retirement based on years of service and final earned income, and promise to protect the real value of retirement income through cost-of-living adjustment. Should the plan not be able to fulfill its promises, the burden passes to the employer. If the employer by reason of bankruptcy is unable to meet those obligations, the UK government's Pension Protection Fund (PPF) is there to provide a formula-based discounted value of those benefits. By contrast, the value of a DC pension is the product of employer and employee contributions and the accumulated investment returns on those assets. No promises are made about the final value of a DC pension; the employer does not stand behind the real value of the accumulated assets, and no commitment is made to the income of plan participants through their retirement years. The employee bears the risks of pension saving, including whether the accumulated pension is adequate through retirement or not.

In the years following the Second World War, private employers, prompted by unions and government legislation outlawing certain types of workplace discrimination, effectively broadened the DB franchise, taking in large and small companies as well as high-paid and less well-paid employees (Clark, 2000). Whether directly or by default, employees were enrolled into DB pension plans through collective bargaining agreements and employment contracts; in many sectors, workers had neither the option not to enroll in the company pension scheme nor a choice about what level of contribution to make to the scheme from weekly or monthly salaries. Enrollment, contribution rates, and eligibility were set by the scheme. By contrast, UK and US employees are typically given the option to enroll or not enroll in DC schemes, as well as some choice over the level of contributions to make from current incomes. As is widely noted, low-waged workers tend not to enroll in such schemes, and, if they do so, they tend to choose low rates of contributions (compared to higher-waged employees) (Benartzi and Thaler, 2001).

When enrolling in a DC pension scheme, participants are often required to make decisions about the allocation of contributions to different asset classes and, within those asset classes, decisions about investment products and service providers. As is widely noted, however, the younger a plan participant, the less likely he or she is to reflect upon the significance of these decisions and the virtues or otherwise of alternative courses of action. The available evidence suggests that participants are quick to decide and are slow to change their initial allocations (see Iyengar et al., 2004; Madrian and Shea, 2000). Intuition and inertia dominate unless the plan sponsor provides information, decision tools relevant to participants' age, income, and gender, and the means by which participants can cut through the available choices to points of reference that are salient to the circumstances of participants (Chapter 4). In some jurisdictions, plan providers are permitted to offer so-called default settings such that, if participants are unwilling or unable to make such choices, they are automatically enrolled in a common investment fund.

DC pension plan participants are required to make decisions in the context of market risk and uncertainty. As such, the lessons of the behavioral revolution allied with Herbert Simon and Kahneman and Tversky (1979) apply with some force; that is, identified behavioral biases and anomalies can be shown to be especially relevant to the DC domain. For example, the fact that many people are short-term oriented when in fact they are required in the DC domain to act in terms of their long-term interests, and that many people are quite myopic when in fact they are required in the DC domain to judge the significance of events in relation to the underlying process whereby pension contributions cumulate to retirement, suggest that, when left to themselves, most people lack the information and skill to plan for an adequate retirement income (Strauss, 2009*a*). It is also obvious that many people are quite naïve when it comes to making decisions in the DC domain where, in fact, a high level of sophistication is required by virtue of the uncertainties of financial markets. Since professional portfolio managers are, on average, unable to systematically "beat the market," this gives some indication of the challenges that individuals face when acting as portfolio managers in their own right.

This has been recognized by behavioral theorists and pension specialists (Thaler and Sunstein, 2008). Nonetheless, the design and structure of many DC schemes remains firmly wedded to the ideal of the rational actor. This has implications for retirement planning and saving (Lusardi and Mitchell, 2007).

Retirement Saving and Housing

It seems likely that many of those who rely upon DC occupational pensions to meet their retirement income aspirations face the prospect of those aspirations not being realized. In part, this may be because DC savings rates are typically too low and because many people simply do not appreciate the difference between DB and DC pension schemes including the significance of employer contributions (Chapter 3). For those who do come to terms with the discounted value of DC plans, one implication is that other forms of saving may be needed to insure against not realizing retirement income aspirations. The burden of assessing any shortfall in the expected value of an occupational pension and any response to anticipated shortfalls in value are clearly the responsibility of individuals rather than institutions.

Just as the government has discounted the future value of state pension entitlements, so employers have effectively discounted their future commitment to occupational pensions. One implication is that the third pillar of retirement savings will have to compensate for the discounted values of pillars 1 and 2, despite the fact that private pension wealth is extremely unevenly distributed in the United Kingdom (Pensions Commission, 2004*b*). Another implication is that retirement saving should be seen as an exercise in portfolio management, such that reliance on any one element such as the basic state pension or an occupational pension should be balanced against individuals' other options, including savings accounts, insurance, individual savings accounts (ISAs), property, and housing. At a time when academic research has confirmed the limits of human cognition and reasoning in the context of risk and uncertainty, effective planning for future retirement income requires a high level of judgment about balancing the risks of various savings options. Portfolio management is one of the foundations of modern financial theory; it is also applicable to retirement saving, however unlikely it is that most individuals can be effective portfolio managers (Sharpe, 2007).

In previous chapters, we explored a national representative sample of the UK population, drawing inferences about their risk propensities in regard to pension saving according to socio-demographic characteristics as well as measures of risk tolerance. In order to improve our understanding of risk management strategies of individuals in relation to the various saving instruments they may have at hand, we acquired access to the participants of a DC scheme—identified in Chapter 5 as sponsored by a large multinational bank located in London with affiliates around the world. Given the lead-up to the peak of the global financial market bubble, we sought to determine whether participants in this scheme identified property and housing as important savings instruments when compared to their occupational pensions. In other words, we sought to determine whether other savings instruments complemented their current and past pension entitlements or were, in some sense, alternatives to these schemes. In the media, property and housing are touted as alternatives to occupational pensions.

The average DC pension plan participant is likely to be a relatively naïve financial planner and is thus probably unable or unwilling to conceptualize the risks associated with DC pension saving in relation to other savings vehicles. By focusing upon the participants of a plan sponsored by a large financial institution, we had access to a group of participants whose skills and competences are relevant to investment management. In Chapter 5 we showed that there were, in fact, a number of younger, relatively lower-income plan participants who apparently relied heavily upon investment property for their future retirement incomes. We also showed that the majority of plan participants' savings portfolios were biased toward low-risk financial instruments rather than property or equities. Importantly, we were able to show that to the extent that older higher-income plan participants invested in property, they did so in a manner that seemed to imply a quite sophisticated balance between property and other much less risky financial instruments. This suggests that certain groups—women, young people, those on low incomes—who self-report relatively high or low levels of risk tolerance, are in fact those least able to hedge against losses or the risk of inadequate income in old age.

By focusing upon property excluding the family home, we treated the available financial products as a set of complementary instruments that may generate a long-term retirement income. By contrast, the family home is a more complicated entity, combining consumption and investment as well as cultural and social meaning that goes well beyond an instrumental conception of utility maximization (Smith, 2008). Most obviously, property investment such as "buy-to-let" attracts significant tax benefits, whereas the mortgage interest paid on the family home does not. In any event, the family home represents a form of current consumption (shelter) essential to its occupants, while being simultaneously a means by which family members gain geographical access to the workplace, social goods such as health and education, and cultural amenities. In urban Britain, the neighborhood where people live, the type of house they live in, and the social standing of neighbors carry a symbolic, even emotional, value partly capitalized in house prices and partly capitalized in cultural respectability (Munro and Smith, 2008).

As such, the family home is also a long-term commitment to a certain quality and quantity of housing. To the extent that life-cycle issues dominate the consumption of housing, retirement may mean rationalizing the "overconsumption" of housing through equity release in exchange for a flow of income.[3] The family home is an investment in a number of ways. For those concerned about the possible costs of old-age nursing, holding and even enhancing the equity embedded in the family home is a means of self-insurance, thereby ensuring independence from children and the state for as long as possible. For those "house-rich" in relation to needs, the family home can also generate income through the rental of unused rooms. Closely related is the prospect

---

[3] See the Pensions Policy Institute (2009) and JRF (2010) for reports on the virtues of such schemes and Parkinson et al. (2009) on the significance of equity release for Australia and the United Kingdom during the period 1999–2007. Whether equity release is actually translated into a flow of income (annuity) as opposed to simply being consumed is more problematic.

that the family home may be a bequest to surviving family members, should its value be protected from the financial costs of the last years of life. As Poterba (2006) noted, the bequest motive is a significant factor in explaining the resistance of retirees to annuities (see also Chapter 8 in this book).[4]

The home has been treated in the media as an important component of future retirement income. Typically, two types of arguments are made for its importance. For some, the home together with property including "buy-to-let" is believed to be a form of long-term investment that does not share with DC pension investment the ups and downs of financial markets. This was the argument put forward in parts of the British media, bolstered by the popularity of real-estate-themed television programs, through much of the first decade of the twenty-first century. However, as the most recent financial crisis demonstrated, there is actually a close correlation between the performance of property markets, the volatility of financial markets, and the performance of DC pension fund investments. Indeed, Reinhart and Rogoff (2009) argue that discontinuities in property markets often trigger market uncertainty and financial crisis. In this sense, the assumption of "separability" reflects either a lack of understanding about the macro-performance of related markets and/or a lack of understanding about the correlated nature of the risks associated with property investment and DC pension investment.[5]

A second argument is that the risks of property investment, and in particular the home, are more easily understood than the risks associated with DC pension investment because the former are "local" whereas the latter are "global." That is, it is assumed that property and house prices embody significant idiosyncratic factors (location, attributes, and architectural form) such that home owners may be justified in believing they are better able to take advantage of information asymmetries than others not so placed. There is, of course, a significant body of literature in economics and geography that would support the argument that property markets (more than other types of markets) are characterized by opaqueness and high costs of third-party information acquisition (Clark and O'Connor, 1997; Coval and Moskowitz, 2001). Nonetheless, people may

---

[4] This may go some way to explaining why many older people choose not to downsize, even when "house rich" but cash poor. See Banks et al. (2007), Disney et al. (2002), and Venti and Wise (1990).

[5] See also Khandani et al. (2009) on the ways in which the risks associated with US housing purchase and refinancing over the past decade were "vastly" amplified by a variety of factors including "near frictional-less refinancing options," creating an extraordinary situation where many households were exposed to systemic risks in the financial market that they neither understood nor made provision against. Stango and Zinman (2009) provide evidence to the effect that in household finances many people systematically underestimate the costs of borrowing. While focused on the United States, this type of research is clearly relevant to the United Kingdom. See, for example, a recent comment in the Money section of *The Sunday Times* (January 31, 2010) on the pension–property choice, to the effect that property is always preferable because pensions are (seemingly) much more vulnerable to the stock market.

overemphasize the value of their local knowledge and underestimate the systemic global risks associated with property markets; overconfidence is a close correlate of financial naïvety, especially if social status, location, and home ownership are reinforced by the media as valuable cultural capital (see generally Mellers and McGraw, 2004 with Smith et al., 2009).

In these ways, there is a close, even intimate, connection between the risks of property investment, and the home in particular, and the risks associated with pension investment. Here, we are concerned with the risk propensity of relatively sophisticated respondents who indicated that their home was an ingredient in retirement planning. Did they also recognize these overlapping risks and pursue a safety-first strategy, or was a preference for the home as a form of retirement saving associated with risk tolerance? Here, we develop the assessment of risk propensity developed in Chapter 3 with a series of risk measures that go beyond our initial measure.

Survey Design and Risk Measures

As noted in the previous chapters, our survey, implemented at the peak of the bubble (May 2007), targeted more than 7,000 London-based employees of a large international investment bank. The survey was based upon our related national representative survey developed with the project sponsor and augmented with a series of questions that focused upon participants' savings portfolios wherein the employer-sponsored DC plan was identified as one source of saving for the future. A total of thirty-one questions were asked mixing together the collection of personal data with attitudes and prompted solutions to puzzles. A number of questions focused upon the role of property investment in savings portfolios, distinguishing between investment vehicles like buy-to-let and respondents' homes. The full survey is available from the authors.

Most importantly, we sought to clarify how respondents conceptualized the risk of long-term savings strategies when given evidence to the effect that employees/participants were faced by overlapping and reinforcing risks related to their current incomes and future prospects. How people cope with risk in the context of market volatility may provide us with insight about the heterogeneity of behavior as well as the range of strategies deployed by participants to deal with current and expected market conditions. Based on the survey, we have shown that older higher-income participants pursued rather sophisticated "safety-first" property investment strategies, seemingly designed to dampen risks across their savings portfolio. In this chapter, we are particularly concerned with the issue of whether participants viewed their homes as instrumental variables in their retirement savings strategies. In the first instance, we sought to characterize their intentions in this regard according to their socioeconomic characteristics. As well, we sought to distinguish between respondents according to their relative sophistication as retirement planners, focusing upon their levels of expressed confidence, knowledge, and understanding of the issues.

In the second instance, we also sought to test whether those who indicated that they would rely upon the home for their retirement needs had distinctive risk preferences when compared to those who did not, in fact, indicate such a commitment. Specifically, Question 8 asked respondents *"to what extent do you agree or disagree with the following statements"* where six statements were made beginning with *"planning for retirement is an important issue for me"* and the fourth statement which read *"the house that I live in will provide the majority of my retirement needs."* This question came after the introduction to the questionnaire, where a series of seven questions were posed focusing upon age, income, marital status, job classification, and educational qualifications. Following Question 8, we asked a series of questions to characterize respondents' retirement savings portfolios and their knowledge and understanding of retirement planning.

Most importantly, in Section 4 we asked four questions designed to elicit respondents' attitudes to risk in the context of "pensions and long-term savings." Each question related to risk preference is expressed somewhat differently and is based upon a particular conceptualization of risk. For example, as Question 25 described in Chapter 3, this Question 17 asked: *"When you are thinking about long-term savings and pensions, which of the following summarises your attitude: I aim to get the best possible growth in the value of my savings, even if that means taking the chance that I could lose money; or I prefer to have safe and secure savings and investments, even if that means they do not grow in value as much as they could."* Here, respondents face a seemingly simple choice between maximizing the risk-adjusted rate of return and pursuing a safety-first strategy albeit with a lower rate of return over the long term. This question represents Roy's argument (1952: 432) to the effect that "real people" do not have "precise knowledge of all possible outcomes of a given line of action, together with their respective probabilities" and are risk averse especially as regards the possibility that an economic disaster might wipe out their savings.

Questions 18 and 19 test whether respondents are risk averse in the manner suggested by Kahneman and Tversky's prospect theory (1979). These two questions have much the same logic but focus separately on the components of their value function. In Question 18, risk aversion is tested by reference to the certain receipt of a sum of money against the probability of winning a larger amount, whereas in Question 19 loss aversion is tested with the certain loss of a sum of money against the probability of losing more. Specifically, Question 18 asked: *"If you were given the following choice, which option would you prefer? Please select one: receiving £3000 with certainty or a four in five (80%)*

chance of winning £4000." This is followed in Question 19 with: *"If you were also given the following choice, which option would you prefer? Please select one: losing £3000 with certainty or a four out of five (80%) chance of losing £4000."* See Baron (2008: 277) on the logic of prospect theory and Trepel et al. (2005) on the different weights attributed to the components of prospect theory.

Notice, in contrast to Question 17, in Questions 18 and 19 it is assumed that respondents are able to judge the probability of certain outcomes distinguishing between certainty on the upside and a possible downside. In this case, it is reasonable to assume that respondents were able to deal with chance expressed in ratios and percentages since making such judgments is important to their everyday working lives in the finance industry. As has been shown elsewhere, this assumption of computational competence is not entirely plausible if considered in relation to the wider population (Oaksford and Chater, 2007).

A less sophisticated test of respondents' risk propensities was set out in Question 20: *"When making decisions about your own finances are you prepared, if rewards could be large, to accept: a large degree of risk; a moderate degree of risk; a small degree of risk; a very small degree of risk, or I don't know?"* Here, we assumed people can and do make a distinction between how they would respond to different levels of risk. In a sense, this question tests whether people respond to risk in relation to the proffered reward or whether they, in fact, prefer certainty in their own finances if not in their jobs in the financial sector. Given these measures of risk, we also sought to test whether respondents were consistent in their responses and whether consistency was an indicator of respondents' likelihood of indicating that their home would be an element in retirement planning. Consistency of response is rarer than might be imagined: people are, more often than not, seduced by the framing of questions and tend not to appreciate the commonalities between supposedly different tests of risk preference (Krueger and Funder, 2004).

Statistical Analysis and Results

Ordinal logistic regression models were developed to explain variation in attitudes to the statement *"The house that I live in will provide the majority of my retirement needs."* As noted in previous chapters and by Hills (2010) among many others, the following socio-demographic variables are important in distinguishing between individuals' attitudes and options as regards retirement planning:

- Gender
- Age

- Education (higher education; possession of professional qualification)
- Job type (level of responsibility)
- Marital status (including "partnership")
- If married or cohabiting, whether their partner had an occupational pension
- Base salary (excluding bonuses).

Cross-tabulation of the data revealed low or zero cell counts for some combinations of levels of the dependent variable with levels of explanatory variables. To ensure consistent parameter estimates, some levels of these factors (age, base salary, and the dependent variable) were combined to ensure adequate cell counts.[6] The data on partner's occupational pension were also recoded to a binary variable with levels "is/is not married or cohabiting with a partner with an occupational pension" in view of the minority of positive responses from respondents in the single and separated/divorced/widowed categories. An overview of the modified dataset can be found in the published paper.

Candidate models were fitted under a proportional odds assumption using the R and SPSS statistical packages. A stepwise selection algorithm minimizing AIC (Akaike information criterion) was used to find candidate models, with a model including only a constant used as a starting point in view of the sparse dataset relative to the number of predictors.[7] The resulting model is presented in Table 6.1. Both the possession of professional qualifications and the pension status of partners are not included in the selected model minimizing AIC. The remaining explanatory variables are each significant at the 1 percent significance level or above/under a drop-in-deviance test. The coefficients for the levels of these factors, together with corresponding standard errors and estimated odds ratios, are also to be found in Table 6.1.

The model is formulated such that the interpretation of the coefficients is as follows: e^{β_i} is the ratio of the odds for someone who falls into the set indicated (e.g., women, those with higher education degrees) of being in a group which agrees more with the statement *"The house that I live in will provide the majority of my retirement needs"* relative to the corresponding odds for the reference group, ceteris paribus. The reference group is set to include males of age 29 years or younger, without a higher education degree, of non-officer job type, single, and with base salary less than £25,000. For example, considering possession of higher education, an estimated coefficient of less than 1.0

---

[6] For age, levels {50–59 years}, {60–65 years}, and {over 65 years} were merged into a single category; for base salary, levels {£15,000 or less} and {£15,001–£25,000} were merged; and for attitudes to the statement "The house that I live in will provide the majority of my retirement needs," levels {Agree} and {Strongly agree} were merged.

[7] AIC, or Akaike's information criterion, is a measure of goodness of fit which describes the trade-off between bias and variance in model construction, and is thus used as a tool for model selection. When minimizing AIC, models are ranked and the one with the lowest AIC is considered the best fit.

Table 6.1. Results of ordinal logistic regression for attitudes to property as a form of retirement saving

Intercepts				$\hat{a}_j$	$SE(\hat{a}_j)$
"Strongly disagree"|"Disagree" or above				−0.988	0.334
"Disagree" or below|"Neither agree nor disagree" or above				0.974	0.334
"Neither agree nor disagree" or below|"Agree" or above				2.547	0.338

Coefficients	$\hat{\beta}_i$	$SE(\hat{\beta}_i)$	Odds ratio $(\exp(\hat{\beta}_i))$	95% confidence interval for odds ratio	
				Upper	Lower
*Gender****					
Male (baseline)	0.000				
Female	0.131	0.092	1.140	0.952	1.364
*Age****					
29 years or under (baseline)	0.000				
30–39 years	0.379	0.111	1.461	1.176	1.817
40–49 years	0.701	0.142	2.016	1.526	2.666
Over 50 years	0.697	0.206	2.008	1.340	3.008
*Possession of higher education degree (BA, MA, MSc, etc.)***					
No higher education degree (baseline)	0.000				
Higher education degree	−0.286	0.099	0.751	0.619	0.912
*Job type****					
Non-officer (baseline)	0.000				
Associate Director	0.008	0.123	1.008	0.792	1.284
Director	−0.242	0.166	0.785	0.567	1.087
Executive Director	−0.309	0.215	0.734	0.481	1.118
Managing Director	−0.479	0.291	0.619	0.350	1.096
*Marital status****					
Single (baseline)	0.000				
Married/cohabiting	0.314	0.096	1.369	1.136	1.652
Separated/divorced/widowed	0.004	0.253	1.004	0.611	1.646
*Base salary****					
£25,000 or less (baseline)	0.000				
£25,001–£40,000	0.141	0.321	1.151	0.614	2.169
£40,001–£65,000	0.095	0.327	1.100	0.581	2.093
£65,001–£100,000	0.157	0.349	1.170	0.590	2.327
Over £100,000	−0.183	0.386	0.832	0.391	1.778
Residual deviance: 5330.98	AIC: 5366.98	n = 2118			

Strength of significance: ***p< 0.001; **p< 0.01; *p< 0.05; p < 0.1.

indicates that a person who possesses a higher education degree (but otherwise shares the characteristics of the reference group) is less likely to have a positive attitude to the above statement than a person from the reference group.

It can be seen from Table 6.1 that, with increasing age, respondents were more likely to have more positive attitudes to home ownership as a form of retirement saving, with the odds of being in a more positive group increasing by a factor of 2.008 (95 percent confidence interval (CI): 1.340, 3.008) from between the 29 years and under group and the over 50 age group, ceteris paribus. Holding all other variables constant, the odds of being in a more positive group decrease with possession of higher education by a factor of 0.751 (95 percent CI: 0.619, 0.912). The odds of being in a more positive group increase for respondents who were married or cohabiting relative to those who were single by a factor of 1.369 (95 percent CI: 1.136, 1.652). The magnitude of the standard errors for the other coefficients prevents a definitive interpretation of the relationship between the remaining factors and attitudes to home ownership as a form of retirement saving, although each factor itself is significant at a 1 percent level or greater.

Assessment of the adequacy of model fit is complicated by the fitting of the model to a large number of factors, some with several levels. With sparse data, assumptions regarding the approximate distribution of the deviance statistic may not hold (Wrigley, 2002). Hence we examined the adequacy of the model in other ways:

- A likelihood ratio test of the drop in deviance between the saturated model including possession of professional education and partner's pension status and the model presented above found no evidence to reject the simplification to the above model (p-value = 0.88).
- A score test of the proportional odds assumption found no evidence (p-value = 0.24) to reject the null hypothesis that the assumption holds.

To examine the sensitivity of our results to potential sparse data effects, we refitted the model by (*a*) collapsing categories of the response variable (to {Agree}, {Neither agree nor disagree}, {Disagree}) and (*b*) converting age and base salary to continuous variables. In both cases, models including the above six factors only were found to minimize AIC. Furthermore, the magnitude and direction of the coefficients for levels of gender, age, job type, marital status, and possession of higher education were all found to be broadly similar under the different parameters.

With the selected model, we considered four additional variables for inclusion in the model. As noted earlier, three variables were concerned with risk aversion, while the fourth was related to loss aversion (in a manner consistent with Kahneman and Tversky's 1979 prospect theory). Drop-in-deviance tests found extremely strong evidence in each case (p-value $< 1 \times 10^{-4}$) to reject the null hypothesis that the smaller model (excluding the risk variable) was sufficient. The effects on AIC of the addition of each risk variable separately to the above model were considered. In the interests of parsimony, however, only the risk variable with the greatest associated drop in AIC was selected for inclusion in the model—the scaled risk preference measure.

Housing, Retirement Saving, and Risk Aversion

Having tested for the contribution that respondents' risk preferences make to the estimated model, we also considered the extent to which the various risk measures were related—the extent to which respondents could be thought consistent across the risk measures in their conceptualization of risk aversion. Here, we can make a number of observations based upon pairwise comparison. The simplest comparison was made between the scaled risk preference option (from a small degree of risk to a large degree of risk) and the categorical distinction made between a preference for "safe and secure savings" and aiming for the "best possible growth in the value of my savings." It was found that scaled risk preferences were weighted toward a small degree of risk when paired with "safe and secure," and weighted toward moderate to large risk when paired with "best possible growth." Likewise, risk preferences were weighted toward small and moderate risk when paired with the preference for "receiving £3000 with certainty" and weighted toward moderate and large risk when paired with "a 4 in 5 (80%) chance of winning £4000."

However, we could not identify a pattern between scaled risk preferences and loss aversion whether "losing £3,000 with certainty" or "a 4 out of 5 chance of losing £4000," nor could we discern a pattern in responses when pairing "safe and secure" with loss aversion. A chi-squared test of independence of respondent answers to the two questions found no association between responses to "safe and secure" and the loss aversion questions (p-value = 0.502). That is, in these instances it can be said that respondents' scaled risk preferences were unrelated to loss aversion. Nonetheless, when "receiving £3000 with certainty" and "a 4 in 5 (80%) chance of winning £4000" were paired respectively with "safe and secure" and "best possible growth," it was also noted that there was a strong relationship between the first pair but no relationship between the second pair. We also considered whether there was a pattern in responses to the pairing of "receiving £3000 with certainty" and "a 4 in 5 chance of winning £4000" with respectively "losing £3000" and "a 4 out of 5 chance of losing £4000." If respondents were consistent with Kahneman and Tversky's prospect theory (1979), responses would link "receiving £3000" and "a 4 out of 5 chance of losing £4000." Of the 2,240 records used, just 50 percent were consistent with prospect theory.[8]

The results for the refitted model including the scaled risk preference variable (Question 20) are shown in Table 6.2. Each of the variables in the fitted model is significant at a 5 percent level or below. The inclusion of the risk variable results in a drop in AIC of 157 (to 5,210). The odds of being in a more positive group with respect to the home as a form of retirement saving

---

[8] Testing for loss aversion is more problematic than may be thought: see Tom et al. (2007) for an experimental procedure that raises issues as to the testing procedures and proper levels of loss that may be needed to calibrate fully human predisposition to loss aversion.

Saving for Retirement

Table 6.2. Results of ordinal logistic regression for attitudes to property as a form of retirement saving, including attitude to financial risk as a predictor

Intercepts			$\hat{a}_j$	$SE(\hat{a}_j)$
"Strongly disagree"|"Disagree" or above			−1.628	0.384
"Disagree" or below|"Neither agree nor disagree" or above			0.362	0.383
"Neither agree nor disagree" or below|"Agree" or above			1.953	0.385
Coefficients	$\hat{\beta}_i$	$SE(\hat{\beta}_i)$	Odds ratio ($\exp(\hat{\beta}_i)$)	95% confidence interval for odds ratio
				Upper Lower
*Gender***				
Male (baseline)	0.000			
Female	0.036	0.096	1.037	0.860 1.251
*Age****				
29 years or under (baseline)	0.000			
30–39 years	0.390	0.113	1.477	1.184 1.844
40–49 years	0.699	0.145	2.011	1.513 2.676
Over 50 years	0.684	0.209	1.982	1.316 2.985
*Possession of higher education degree (BA, MA, MSc, etc.)**				
No higher education degree (baseline)	0.000			
Higher education degree	−0.250	0.101	0.779	0.638 0.950
*Job type****				
Non-officer (baseline)	0.000			
Associate Director	−0.007	0.125	0.993	0.777 1.270
Director	−0.216	0.169	0.806	0.578 1.123
Executive Director	−0.312	0.218	0.732	0.478 1.122
Managing Director	−0.481	0.294	0.618	0.347 1.100
*Marital status****				
Single (baseline)	0.000			
Married/cohabiting	0.286	0.097	1.331	1.100 1.612
Separated/divorced/widowed	−0.047	0.256	0.954	0.577 1.574
*Base salary****				
£25,000 or less (baseline)	0.000			
£25,001–£40,000	−0.005	0.339	0.995	0.512 1.938
£40,001–£65,000	0.001	0.344	1.001	0.510 1.968
£65,001–£100,000	0.054	0.367	1.056	0.515 2.171
Over £100,000	−0.244	0.402	0.783	0.356 1.725
*Amount of risk prepared to accept when making decisions about own finances, if rewards could be large****				
A very small amount of risk (baseline)	0.000			
A small amount of risk	−0.343	0.181	0.709	0.498 1.011
A moderate amount of risk	−0.556	0.179	0.573	0.404 0.814
A large amount of risk	−1.119	0.239	0.326	0.204 0.522
Residual deviance: 5166.45	AIC: 5210.45	n = 2068		

Strength of significance: ***$p < 0.001$; **$p < 0.01$; *$p < 0.05$; $p < 0.1$.

decrease with an increasing preference for risk. Relative to a respondent prepared to accept only a very small amount of risk when making decisions about their own finances, the odds of being in a more positive group predisposed to the home decrease by a factor of 0.326 (95 percent CI: 0.204, 0.522) for a respondent prepared to accept a large amount of risk, holding all other variables constant.

With the exception of the coefficients for levels of base salary, the magnitude and direction of the estimated parameters remain comparable to those in the model without the included risk variable. Again, with increasing age, respondents were more likely to view the home as providing most of their retirement needs, with the odds of having a more positive attitude to the statement increasing by a factor of 1.982 (95 percent CI: 1.316, 2.985) from the youngest group (29 years and under) to the oldest group (over 50), when holding other variables constant. Possession of a higher education degree was associated with a decrease in the odds of having a more positive attitude to the statement by a factor of 0.779 (95 percent CI: 0.638, 0.950), ceteris paribus, while the odds of agreement for a respondent who was married or cohabiting increased by a factor of 1.331 (95 percent CI: 1.100, 1.612) relative to a single person.

While the size of standard errors for other coefficients prevents a conclusive interpretation, it may be noted that increasing seniority of job type was associated with a decrease in the odds of responding positively to the statement. In contrast, women were more likely to respond positively than men. The effects of base salary on attitudes to the statement were less clear, however, and more subject to variation between fitted models.

Synthesis and Interpretation

In sum, just 250 of the 2,256 usable responses agreed with the statement, leaving 1,429 who either strongly disagreed or disagreed with the statement. Considering that the statement is framed with reference to the "majority of my retirement needs" as opposed to relying exclusively upon the family home, being a relatively open statement with scope for ready agreement at the margin, it would seem that respondents appeared not to fit the profile of many of the celebrities interviewed in *The Sunday Times*. This is surprising given the media attention devoted to the topic in the lead-up to the 2007 financial crisis (and thereafter).

It was found that those who tended to agree/strongly agree with the statement tended to be older rather than younger, were more likely to be married or cohabitating than single, less likely to have a higher education degree, more likely to have a higher income (but not the highest income), and less likely to be senior executives in the company. It is tempting to combine these results into "representative" individuals, notwithstanding the fact that the results hold constant the effects of the other independent variables. Even without doing so, we can make a number of observations based on these results that can help us understand their implications better. For example, it is very likely that as higher-income individuals these respondents may be those who could afford to pursue such a strategy. Likewise, the fact that older respondents tended to agree suggests that they are from the generation that has done well out of the long-term run-up in property and home prices in southern England, enabling reliance upon such a strategy.

From interviews with London-based DC sponsors, it is widely believed that those in the 20–29 and 30–39 age brackets tend to be house-poor, despite their relatively high incomes and bonuses. This may be reflected in another of the variables used to clarify the significant correlates of agreement with the reference statement: a characteristic of the finance industry has been the increasing educational credentials of the average employee and the use of higher education to shift out lower qualified applicants for jobs. This process of shifting has been in train for at least the past twenty years. Finding that those with higher education tended not to agree with the statement could represent the fact that higher education is actually a proxy for being younger rather than older—notwithstanding the inclusion of age in the model, this could represent a generation of employees under pressure to save for a home albeit with a low equity share in any home they "own" (in conjunction with the bank providing the mortgage). Note, however, that in Chapter 5 we showed that these types of respondents tended also to be (more than older, higher paid respondents) investors in property (buy-to-let). In this respect, it is likely that the tax benefits associated with investment property enabled younger, lower-paid respondents to claim a share of the property market but remain relatively underinvested in their home.

Of note is the significance of the measures of risk preference in predicting agreement with the proposition that respondents' homes were important in underwriting retirement income. The stronger the respondents' risk aversion, the more likely they were to indicate, holding constant the socio-demographic independent variables, that they would rely upon the family home for their retirement income. Here, we can accept the hypothesis that respondents tend to believe that the family home is a relatively low-risk investment in their future income. Whether they were justified in that assessment is harder to determine without detailed knowledge of their circumstances before and after the financial crisis.

Since our respondents were likely to be more savvy than the average DC participant, and certainly more aware of the costs of irrational exuberance characteristic of financial bubbles (Shiller, 2000), it is possible that older respondents saw their homes as relatively safe havens in an uncertain world. If so, one implication is that they must have believed that the location-specific value associated with their homes outweighed the possible downside of being caught up in the financial bubble. A less generous interpretation is that they failed to appreciate the ways in which house prices were implicated in the paths of global equity markets and earned income. Alternatively, it is entirely possible that those who responded in the affirmative were simply naïve about the *nature* and *scale* of risks in housing.

The most useful measure of risk preference in terms of its contribution to explanation was the simplest measure: a scaled risk measure rather than a test of reasoning. The next best measure tested respondents' preference for certainty of reward, followed by safety and security, and then certainty of loss. While we were not able to interview respondents to clarify their understanding of the relationship between these measures of risk preference, a preference for low risk would seem to be closely related with a preference for certainty of reward and a preference for a safe and secure return over a risk-related rate of return. Even so, the scaled measure of risk preference would appear to require less deliberation than tests of preference that require respondents to judge the value of alternative courses of action with different potential pay-offs. Unfortunately, we were not able to frame tests of risk preference with potential rewards and possible pay-offs at pay-off levels that our respondents may have found particularly salient to their particular circumstances. It is possible that £3,000 and £4,000 were not taken seriously by our respondents, given their current and expected incomes and the prices of their preference home. Elsewhere, we have shown that some people are quite willing to take a financial gamble on £10,000 and £100,000 considering their age and family circumstances (Clark et al., 2009; see also Tom et al., 2007 on the scale of bet in gambling).

Conclusions

In the aftermath of the 9/11 attack on the Twin Towers, the TMT (technology, media, and telecommunications) bubble burst and Western financial markets recorded two consecutive years of negative returns. Shiller (2000) anticipated the turmoil, arguing that financial markets are subject to recurrent episodes of irrational exuberance unjustified by any rational assessment of the fundamentals. For DC pension plan participants on both sides of the Atlantic, the run-up in financial markets on the back of a promised "new world" of technology had seen account balances grow by as much as 35 percent over the period 1995–2001. However, the TMT bubble exacted a price: many DC pension plan participants saw their account balances shrink by 25 percent over just a couple of years, reinforced by a gloomy assessment of the near future (through to 2005). Yet markets recovered, underwritten by the loose money policies of the Fed and the Bank of England and the leverage applied to credit instruments that promised a "new" era in house ownership and investment returns (Lee et al., 2009). Riding the roller-coaster, many DC pension plan participants recovered their losses and saw their account balances dramatically grow year-on-year through exceptional returns.

When we implemented our survey of DC plan participants, the subprime bubble had carried markets into uncharted territory, just as the subsequent bust carried the global economy into a crisis that threatened to turn the credit debacle into a depression. For our respondents, employed in the finance industry that dominates London, house-price inflation, equity market performance, and the path of earned incomes including bonuses were closely interrelated and highly correlated. And yet, public comment about the respective virtues of property including housing and pension contributions as forms of retirement saving would have had it that these savings instruments are alternatives rather than complements wherein the former is supposed, by some, to be superior to the latter. It is arguable that the pricing of property, and especially housing, has such a significant "local" component that it is reasonable to suppose that one's own home is a means of insuring retirement welfare against the ups and downs of (DC) pension saving.

We have suggested that respondents had the advantage (over the average plan participant) of "being in the market" (reworking Gertler's 2003 argument about the value of tacit knowledge). As such, it is arguable that they were more likely to be sophisticated investors rather than naïve investors, even if opportunistic rather than long-term investors. The first question was about how many of our respondents intended to rely upon the family home for their retirement welfare. Only about 10 percent indicated that this was their intention. The next question was whether those who would do so could be characterized as risk averse, exploiting the suggestion made in the UK media to the effect that property is "more understandable" than pension savings. There is strong evidence to the effect that those who expressed an intention to rely upon the family home for a "majority of their retirement needs" were highly risk averse, reinforcing our previous findings on gender differences in risk tolerance.

When we looked for socio-demographic predictors of this preference, it was noted that older rather than younger employees, higher rather than lower-income people, and married/with partner rather than single/separated were more disposed to indicate reliance upon the family home for the "majority" of their retirement needs. Based upon our research that sought to characterize the planning intentions of a representative sample of UK pension plan participants, it would seem that the salience of the issue (future retirement) and the available asset (given their incomes and marital status) were likely to be relevant factors in "explaining" these findings. Put in the negative, younger, lower-income, either never married or divorced people were less likely to agree with the statement, suggesting a lack of salience of the issue relative to other considerations, as well as a less significant claim on such an asset (home).

These findings highlight the fact that many people may not have a significant claim on such a resource and, as such, may be less well off than others so endowed. As Hills (2010) has noted, UK disparities in wealth and earned income are growing, with significant implications for the retirement welfare of the baby-boom generation and, especially, the generations to come.

Of course, given the small numbers involved, it is entirely possible that those who indicated reliance upon their home did so because they had been the beneficiaries of the acceleration in house prices over the 2001–7 period (and for older participants, earlier periods of house price inflation), and they expected that, whatever the costs of the looming "correction," their home would be insulated from the downturn in the market. Perhaps by virtue of age, income, and status, they lived in London neighborhoods that had effectively weathered past financial turmoil compared to the ups and downs of pension balances—they were, or are, actually very savvy investors. Equally, it is possible that they treated the family home as a low-risk investment for their future well-being while treating their pension account as a risky option on the future value of global stock markets. In other words, respondents may have believed that their pension account was subject to greater market volatility than the value of their home, given its location, and that, while the upside of financial markets is highly correlated with house prices, the downside of markets is not so correlated for their *particular home and its neighborhood*. In this sense, our respondents may have been risk averse, balancing the risks found in their retirement portfolio.

If plausible, these types of sophisticated retirement planners do not appear in the Money section of *The Sunday Times*. Rather, it seems that they have strong views justified, more often than not, by simple statements of belief and past success and failures. More research is needed to understand how people manage the risks of different savings instruments, separately and together, placing them in context—that is, in the world in which they live with its various advantages and disadvantages by virtue of who they are and where they live (Strauss, 2009*b*).

7

The Demand for Annuities

On January 14, 2000, the Dow Jones Industrial Average (representing the performance of blue-chip companies on the New York Stock Exchange) reached what was then counted as an all-time high of 11,722.98 points (BBC, 2003). It was widely believed that this heralded a new era in global economic prosperity based upon the technology, media, and telecommunications (TMT) sectors located in global leading-edge clusters of innovation like Silicon Valley. Notwithstanding the media hype and market frenzy, some analysts were highly skeptical (Shiller, 2000). In the aftermath of the Enron and WorldCom governance scandals and the September 11, 2001 attacks on the World Trade Centre in New York, the developed world watched the bubble burst with a mixture of shock, fear, and resignation. Over the next couple of years, those reliant upon stock markets for their retirement income saw the value of their accumulated savings fall dramatically. By October 2002, the Dow Jones had declined to 7,286 points with the FTSE100 bottoming out at 3287.00 points on March 12, 2003 (BBC, 2003).

As explained in previous chapters, the well-being of pensioners and the performance of stock markets have become, for better or for worse, intimately entwined (Langley, 2008; Minns, 2001). There is a premium on people's anticipation of, planning for, and responses to the vagaries of financial markets; as we have seen, such behavior requires finely calibrated judgment and decision-making under conditions of risk and uncertainty (see Chapter 4). To be most effective, people should be able to judge the significance of current circumstances against their long-term interests, a process of accommodation that depends upon a sophisticated knowledge and understanding of financial markets (Chapter 5). As we have shown in previous chapters, financial and pension risk-related decision-making is to some extent associated with socio-demographic characteristics including age, gender, income, and marital status (see, e.g., Chapter 3).

The first bubble and bust of the twenty-first century was an extraordinary period of time. When viewed against the twentieth century, only the 1929

The Demand for Annuities

stock market crash and the 1930s depression were thought to be as significant (see Lowenstein, 2004). But, of course, they were a precursor to the global financial crisis. In an increasingly integrated world, subject to the flux and flow of financial forces in even the most remote corners of the globe (Barro, 2006), it is arguable that the TMT and subprime bubbles are representative of a level of market volatility that is here to stay (Borio, 2006). Recognizing that financial planning, including retirement saving, is, more often than not, highly responsive to immediate circumstances, the planning issue appears far more complex than perhaps hitherto appreciated. Even if less studied, we suggest in this chapter that there may be also a significant geographical element in individual planning responses to global financial market volatility. In doing so, we follow the lead of Gigerenzer (2004) and others in suggesting that behavior (financial and otherwise) is framed by an individual's cognitive abilities, their social identities, and their specific circumstances. To think otherwise would be to strip individuals bare of their place in the world.

Drawing upon a national survey of more than 4,500 men and women aged 50–64, the focus of this chapter is on their pension-planning and decision-making three years after the early 2000s crash. After a review of the literature stressing the intersection between recent advances in behavioral finance and our commitment to a context-dependent view of behavior, we describe the unique Towers Watson dataset on "scary markets" which underpins our empirical analysis. The analysis proceeds in two stages, beginning with a test to determine whether the map of UK supplementary pension benefits is correlated with respondents' region of residence or socio-demographic status or both. It is shown that region drops out as a statistically significant factor once social status is included, emphasizing the fact that access to UK supplementary pension benefits is strongly correlated with age, gender, and occupation.

We then turn to whether respondents would take an annuity upon retirement.[1] Again there is a test for the significance of region and social status, and we find that region of residence accounts for much of the observed variance in responses. Notwithstanding the downturn in financial markets, respondents from the South of England were likely to have purchased an annuity, whereas respondents from the North of England, Scotland, and Wales were not likely to have purchased an annuity. These results have significant implications for the map of UK pension welfare (see Sunley, 2000), just as there are significant

---

[1] An annuity is a financial product that returns a set yearly income over a designated period of time (most likely, to the end of a recipient's life) in exchange for an initial nonrefundable sum of money (Poterba, 2006). It is most often used when converting work-life savings into a predictable and secure stream of income through retirement and is, as well, associated with the decumulation phase of defined contribution retirement savings plans (Cannon and Tonks, 2005).

implications for UK pension policy-making and especially the design of the National Employment Savings Trust (NEST) (see Chapter 8).

Pensions and Financial Literacy

The Pensions Commission (2005) (the Turner Report hereafter) was charged with charting the dimensions of the UK pensions crisis and setting an agenda for reform. While the UK pensions system is complex, it can be summarized as comprised of the following: a modest state basic pension, tax-preferred personal and employer-sponsored supplementary pensions, and various mechanisms of individual retirement saving some of which attract a tax benefit, others of which do not. While the Labour government sought to increase the value of the state pension over the past decade, it remains modest in value compared to continental European pension systems that stress income replacement rather than basic need (De Deken et al., 2006). Personal and occupational pensions have been very important for supplementing the shortfall in the state pension, especially for those recently retired and for the baby-boom generation about to retire over the coming decade or so. It is doubtful, however, whether occupational pensions will be as important for many younger workers in the private sector over the next twenty-five years (Clark, 2006*a*).

Many of the baby-boom generation are entitled to a final salary (defined benefit, DB) pension, wherein the plan sponsor shoulders the responsibility of asset allocation and investment risk. These types of plans were commonly associated with large, unionized employers in manufacturing industries as well as the professional classes in the public and private sectors. Most importantly, DB plans were associated with a form of corporate capitalism increasingly vulnerable to global competition and the depredations of the market for corporate control (Clark and Monk, 2007). Most UK DB plans are now closed to new entrants; defined contribution (DC) schemes are growing in importance after a slow start (compared to the United States; see Munnell, 2006), and in some cases employers have retreated from offering any pension benefits at all. Given the challenges facing private sector employers, the Turner Report recommended the establishment of a National Pension Savings Scheme (NPSS)—a government-sponsored, national, multi-employer DC pension scheme based on matching contributions (now termed NEST).[2]

It is arguable that DB pensions are *ipso facto* preferable to DC pensions in that the benefit risk is borne by the plan sponsor. But it is not entirely obvious

---

[2] See the report by the Department for Work and Pensions (2006) sketching the likely design and operational structure of the NPSS, with implications for the provision of services within the NPSS structure and for the interface of the NPSS with existing pension policies, including annuities.

that DB pension promises are full income guarantees (Clark and Monk, 2008). With the establishment of the Pension Protection Fund (PPF), the UK government has underwritten a quasi-insurance scheme wherein pension beneficiaries reliant upon failing companies may be able to claim a discounted portion of promised pension benefits. By this policy, DB pensions are contingent promises in a world of global competition, capital market integration, and limited government resources. Likewise, it is widely recognized that the final value of a DC pension is dependent upon a string of participant decisions including contribution rates, asset allocation strategies, the choice of cost-efficient service providers, and the state of financial markets when converting accumulated assets into annuities. Barring a radical realignment of the public/private pension divide, one way or another, individuals will carry larger not smaller pension risks and greater responsibility for their retirement income over the coming decades (Clark and Whiteside, 2003).

We have argued in previous chapters that the future of supplementary pensions is symptomatic of the increasing dependence of UK residents on markets for many kinds of social goods. So, for example, over the past thirty years much of the public housing stock has been privatized, transforming tenants into property owners. Similarly, the privatization of utilities, transportation, and even parts of the education and healthcare sectors has effectively placed a price on a bundle of goods previously believed to be the responsibility of the state. Not all UK residents are so deeply embedded in markets; there remain sections and regions of society that live outside markets for housing, labor, and transport. But it is apparent that they do so sheltered by remnants of the welfare state while living in considerable deprivation and geographical, social, and economic isolation (see Smith and Easterlow, 2005).

For many people, long-term retirement welfare is a function of their market-based decision-making. As such, governments have sought to improve financial literacy, assuming that improving the quality and quantity of market-dependent knowledge can enhance individual well-being, discipline the market providers of social goods, and encourage individuals to take responsibility for the consumption of financial services (Strauss, 2008*a*). The implications of this policy regime for social welfare have been widely noted in human geography and the social sciences more generally (see, e.g., Easterlow and Smith, 2004; Leyshon et al., 1998). This policy regime has been very important for private pensions. For example, the Thatcher government encouraged workers to opt out of occupational pension plans and purchase personal pensions directly from the financial sector. While considerable effort was put into financial literacy programs by the government, the Turner Report can be read as a response to the pensions mis-selling scandal of the 1990s.

Risk, Decision-Making, and Context

Do people act consistently in accordance with their long-term retirement welfare? The answer provided by standard economic theory is *yes*, in the core proposition underpinning postwar social science; but we have suggested that it is difficult to justify this in the light of research suggesting the existence of systematic biases or anomalies in human reasoning under conditions of risk and uncertainty. We have rehearsed the main arguments for and against the rational actor model in previous chapters: the behavioral evidence suggests that decision-making is bounded, task-specific, framed or anchored by aspirations, subject to the nature and value of information, and less than perfect (Baron, 2008). It has been shown that many people are risk averse, are inadequate users of information, and do not understand the nature and scope of the risks assumed. For the United Kingdom, we have shown that many people are unable to distinguish between different types of pensions (DB, DC, and personal pensions) on these specific grounds (Chapter 3).

Even so, there are nuances in this research program. Doherty (2003) distinguished between optimists and pessimists. He classed Gigerenzer and colleagues as optimists by virtue of their emphasis on the use of effective (if not efficient) heuristics and shortcuts in decision-making, while Kahneman and Tversky were classed as pessimists by virtue of their emphasis on the shortcomings of human reasoning. In this chapter, we align our research with those who dispute the affirmative proposition ("strong" rationality once and for all). We also align our research with the optimists with two provisos. First, as people use heuristics, they may benefit from task-relevant education, professional qualifications, and training (Clark et al., 2007). Second, in this respect, judgment and decision-making cannot be divorced from their context. In fact, a person's "position" can be an asset or a liability in that the resources associated with social position can affect the nature and sophistication of judgment and decision-making (Clark and Strauss, 2008; Strauss, 2008*a*). We have also shown that household assets like a spousal pension entitlement can encourage household members to pool risks and be less risk averse than others not so strongly placed (Chapter 4). Equally, a person's social position may be so short of prospects and resources that the pessimists' expectations are realized.

In a test of the relationship between socio-demographic status and financial decision-making, Hallahan et al. (2004) found gender, age, number of dependents, marital status, income, and wealth to be related to measured individual risk tolerance. Age (being older) and marital status (being married) are particularly significant determinants of risk tolerance, just as gender is correlated with risk aversion where, holding constant age and marital status, women were found to be less risk-tolerant than men. For income and education, any increase in risk tolerance is particularly apparent at the higher ends of the education and income distributions. These findings suggest that emphasizing human shortfalls (as in many experimental situations) may be misleading in the face of evidence suggesting that the degree of risk tolerance is affected by socio-demographic status (Wiseman and Levin, 1996). Hallahan and colleagues' results suggest that, as the population ages and as the gender composition shifts toward women, assets could, all things being equal, shift toward less risky assets, which would have significant ramifications for long-term welfare.

Citing Hallahan et al. implies that their Australian findings are relevant to the United Kingdom. However, the Australian pension system is quite distinctive—being heavily reliant upon mandatory DC supplementary pensions. Country-specific expectations and institutional arrangements can give rise to different patterns of risk tolerance and behavior, holding constant socio-demographic characteristics (Jacobs-Lawson and Hershey, 2005). So, for example, Henrich et al. (2005) in a large cross-national study found that the conventional notion that "individuals seek to maximize their own material gains" whatever the context could not be substantiated. While there was evidence of institution-specific incentives as well as shared cross-national socio-demographic effects, the "local" culture remained a significant effect.[3] Similarly, in a large-scale study of the South African credit market, Bertrand et al. (2005) found that even the "local" context in which respondents deal with risk and uncertainty could produce significant differences between nominally the same people (by socio-demographic measures).

Crucial for the study of human behavior is the issue of contingent behavior: the extent to which unanticipated events produce individual responses at variance with their long-term interests. In the literature, this issue is treated as an instance of "weakness of will," suggesting that those who do not stay the course—being opportunistic, easily distracted, and/or wholly focused on short-term costs and benefits—are irrational or worse. However, this type of behavior may be thought reasonable if not consistent with long-term utility maximization (Ainslie, 2001). Where people are unable to judge the significance of an event in relation to long-term plans, where people suffer an adverse impact from a long-term commitment through an unexpected event, and where people are unable to compensate for such a loss over the foreseeable

---

[3] To illustrate, it is widely believed that, compared to US investors, German investors came late to the TMT boom, overinvested in bonds in the aftermath of the bubble, and did not return to equity markets at the same rate as US investors, even when markets had largely recovered their past positions. It is commonly observed that German investors are relatively risk averse, their attitudes being anchored in past experience and certain cultural predispositions. Behavioral scientists commonly acknowledge the significance of anchor effects in framing the response of individuals to risk and uncertainty (Kahneman, 2003). Unexplained, however, are the long-term cultural and historical foundations of those effects.

future, it would seem reasonable to respond to these circumstances rather than hold true to a receding objective. This type of behavior may be accentuated by a lack of knowledge of the processes at work, and a lack of experience in calculating the odds of the short-term effect persisting into the future.

We would contend that knowledge and experience can make a difference to the exercise of financial judgment under risk and uncertainty (the product of our research on the impact of education and task-specific professional experience on the consistency of decision-making; see Clark et al., 2007). Similarly, the optimists would argue that people's responses to common circumstances should be highly differentiated. The intersection between cognitive capacity and social position is the point where heuristics (strategies of response) are conceived and implemented (a version of Simon's 1956 scissors). But social position, like the term context, carries with it an enormous variety of meanings (as indicated in Chapter 2). Here, we distinguish social position by reference to respondents' socio-demographic characteristics and their region of residence. At issue is the relative significance of each in explaining the observed variance in risk management strategies.

Empirical Strategy

By this logic, the context of behavior could be a distinctive map in time and space (implied by Kahneman and Tversky's 1979 notions of framing and anchoring but rarely articulated as such). This is important for our project for two reasons. It provides the rationale for the study of individual responses to the downturn in global financial markets, just as it provides the rationale for introducing a geographical element into the study of those responses alongside the analysis of the effects of socio-demographic status. As well, this logic provides a rationale as to why we should expect the TMT bubble to have amplified the existing advantages of higher socio-demographic groups located in core regions of the economy. In this section, we explain the empirical strategy used in the project, including the data sources and the tests of significance of both geography and socio-demographic status.

Before doing so, it is important to acknowledge the subjects of analysis: the pension risk-related management strategies of 50–64-year-olds. The evidence suggests that the market downturn was of great significance compared against the experience of financial markets over the twentieth century (Lowenstein, 2004). While we were not able to test the behavior of our respondents after the market crash against their previous behavior, Chapters 3 and 4 provide benchmarks for anticipating behavioral responses. Although willingness (or otherwise) to take an annuity may reflect ingrained behavior, the downturn produced a significant decline in interest rates and bond yields, affecting those committed or required by regulation or necessity to take out an annuity. Annuity rates halved over the period 2002–5, reinforcing the decline of annuity rates over the past thirty years as inflation has declined (see Cannon and Tonks, 2004).[4] Even if individual responses to market expectations are time-contingent and may not be replicated from one time to the next, the logic of analysis may provide clues as to the means of assessing those responses.

In order to analyze the intended responses to the crash, we used data collected by the leading UK commercial polling company YouGov for a "scary market" commissioned by the global financial services firm Towers Watson. The individuals surveyed for the scary market survey were selected from a population of approximately 40,000 respondents (YouGov, 2006a, 2006b). The online scary market survey was sent to a group of individuals in the YouGov panel survey between the ages of 50 and 64, selected because they were nearing retirement or were retired (Gardner and Orszag, 2004). From this group, 4,538 individuals responded.[5] The survey asked individuals how their retirement savings had been affected by changes in equity markets and how those changes had affected their retirement plans. In addition, the survey gathered data on respondents' pension arrangements, income, education, and various other socio-demographic characteristics, including region of residence.

The initial analysis focused upon respondents' number of supplementary pensions (1, 2, 3, 4+) before turning to their willingness to take an annuity over a lump sum at ages 60, 65, 70, 75, or never.[6] A subsample of 2,506 respondents was created for individuals who provided a response to all of the socio-demographic and regional characteristics of interest. Of this subsample,

---

[4] There has been some debate over the efficiency of the UK annuity market; the balance of evidence suggests that UK markets perform relatively "satisfactorily" even if market concentration increased substantially over the past decade (Cannon and Tonks, 2004; see also Murthi et al., 1999).

[5] The YouGov sample may contain bias because of the nature and conduct of the survey. YouGov actively recruits individuals from all ages, socioeconomic groups, and regions of Britain to generate representative population samples, and utilizes a reward system for participation. Nonetheless, as with any survey, respondents are self-selecting and may participate because they are particularly opinionated about a topic. The requirement that individuals have internet access may also introduce bias by eliminating important sectors of the population, particularly in older age and lower income groups who may not have internet access (ONS, 2006c). While the coverage of older people who are less computer literate may be problematic, the sample does have adequate representation (25 percent) of individuals from lower income brackets (earning less than £20,000 per year). Furthermore, to minimize bias all correlations were conducted using YouGov compensatory weighting, which accounts for the difference between the sample profile and actual population structure in terms of age and gender.

[6] At the time of the YouGov survey, the UK government required those participating in tax-preferred retirement savings accounts to convert accumulated assets into an annuity by the age of 75. This policy was the subject of widespread dissention, with some arguing against the policy on the grounds that it disenfranchises participants from leaving a bequest to surviving family members. In early 2006, the UK government changed the annuity policy to allow for the direct draw-down of pension assets and their transfer between generations at death, only to announce at the end of 2006 the imposition of significant tax penalties on such transfers in line with Treasury policy on inheritance.

438 (17 percent) held zero supplementary pensions, 1,120 (45 percent) held one such pension, 605 (24 percent) held two pensions, 225 (9 percent) held three pensions, and 118 (7 percent) held four pensions or more. The number of pension entitlements held can be thought to depend on a variety of factors, including occupation, length of job tenure, and the strength of the local labor market (a strong market would, presumably, encourage job switching and hence the take-up of more supplementary pension entitlements over a person's work life). Notice, however, that the vast majority of respondents indicated that they held just one or two such entitlements. The first step of analysis was to map the UK pattern of pension entitlement by social status.

In terms of respondents' willingness to take an annuity rather than a lump sum, responses were initially separated into one of three categories: those who would always take an annuity, who would never take an annuity, and who would sometimes take an annuity (a mixture of would take, and would not take responses). Of the respondents, 234 (9 percent) would always take an annuity, 1,249 (50 percent) would never take an annuity, and 1,015 (41 percent) would only sometimes take an annuity.[7] This suggested that half of respondents would never take an annuity at any age. A subsequent test of the correlates of the take-up of an annuity distinguished between those who would switch intended behavior by age from would not (60 and 65 years of age) to would (70 and 75 years of age) and would (60 and 65 years of age) to would not (70 and 75 years of age). Having established whether the number of pension entitlements held by a respondent could be said to correlate with social status and/or geography, we then sought to determine whether the take-up of an annuity in the aftermath of the TMT bubble could be correlated with social status and/or geography. Therefore, two sets of empirical analysis were conducted, and both relied upon common socio-demographic characteristics and regions of residence.[8]

Correlates of the Number of Pensions

Two models were used to test the relationship between the number of supplementary pensions held by individuals, their region of residence, and their

---

[7] The ambivalence shown by our respondents for an annuity is consistent with the low voluntary take-up rates apparent in the United Kingdom and the United States (see Cannon and Tonks, 2005: 3–4). Nonetheless, the available academic research suggests that annuities are a valuable means of securing long-term retirement income, especially when longevity is on average increasing, as is the likelihood of older-age poverty becoming a significant problem in many Western economies (see MacKenzie, 2006 on the market for annuities).

[8] Regions of residence began with a set of fourteen such regions, mostly larger metropolitan regions plus the South-west of England, Wales, and a couple of Scottish regions. In the end, these were collapsed into just six macro-regions (the South, the North, Scotland, Wales, the South-west, and London) so as to ensure adequate data in each cell.

The Demand for Annuities

socio-demographic characteristics. The first model assessed the extent to which region alone explained variation in respondents' number of pensions. The second model combined region with socio-demographic characteristics as a test of their relative significance. The dependent variable used in both models was the number of pensions held by respondents. Region, age, gender, marital status, home ownership, children, work status, income, and education were used as explanatory variables in the second model. These variables are described in more detail in the Appendix. Both models were estimated using ordinary least squares (OLS) linear regression. Heteroskedasticity (differing variance) was not found to be a problem in the models because of robust standard errors. The models were estimated using compensatory YouGov weighting to control for age–gender differences in the sample from the UK population.[9]

The results of our analysis are presented in the two regression models described in detail in Clark and Knox-Hayes (2007). It was shown that region of residence was a statistically significant predictor of the number of pensions held: the likelihood of holding more supplementary pensions varies between regions when compared with the baseline of London. Only the South was not statistically significant, a possible indication of its closeness to London. Individuals in London were found to be more likely to hold a large number of pensions than individuals from any other region. Respondents from Wales and the North were likely to hold fewer pensions (with coefficients of -0.14 and -0.15, respectively). Respondents from the South-west and Scotland (coefficients of -0.28 and -0.27, respectively) had the least likelihood of holding a large number of pensions. These results were significant until socio-demographic variables were added into the model.

With the addition of socio-demographic variables, however, the region of residence dropped out of the analysis. Nearly all of the socio-demographic variables were significant predictors of the number of pensions held by respondents. Respondent age slightly increased the likelihood of holding more pensions. Women and individuals who cared for children were found to be significantly less likely to hold a large number of pensions. Not surprisingly, individuals who had higher household income, higher levels of education, were single, and who owned homes were found to be more likely to have a larger number of supplementary pensions. By contrast, those who worked part-time, were retired, and/or were unemployed were found to be less likely to hold supplementary pensions—a reflection, no doubt, of the predicament

---

[9] The YouGov weighting scale slightly "downweights" the responses of men and individuals between the ages of 55 and 59, and slightly "upweights" the responses of women and individuals over the age of 60. The weighting accounts for the fact that women comprise only 30 percent of the sample but close to 52 percent of the UK population. Weighting likewise accounts for the fact that individuals over the age of 60 are slightly underrepresented in the sample, whereas individuals between the ages of 55 and 59 are slightly overrepresented.

facing all lower-income workers. Having a larger number of pensions may be an indication of robust and flexible labor markets in national professional and financial services industries and, consequently, of a sequence of employers over the course of an individual's working life.

Taken together, our findings suggest that men with higher income, who are single and do not have children, hold the most supplementary pensions and thus have the greatest retirement income security. Women, particularly those who care for children and who do not work full-time, are more likely to have the fewest pensions and perhaps the least pension security. In line with our previous findings, these results suggest that, in terms of pension policy and pension plan design, attention should be directed toward boosting the pension security of lower-income women who may have spent time raising children and who have not accumulated significant pension entitlement over their working lives (Bajtelsmit and Jianakoplos, 2000). These points are made in a number of recent commentaries on UK pension welfare (see Pemberton et al., 2006) and empirical analyses of pension security by socio-demographic status, all of which underpin the Department for Work and Pensions (2006) report on the likely structure and organization of the NPSS.

Demand for Annuities

In the first instance, two models were used to test the relationship between respondents' willingness to take an annuity and the significance of their region of residence and socio-demographic characteristics. Two dependent variables were tested in separate models. The first dependent variable, "would always take an annuity," was binary with values equal to 1 when a person in every instance chose an annuity over a fixed sum, and with values set equal to 0 when a person indicated that in no or only some instances would they choose an annuity over a fixed sum. The second dependent variable, "would never take an annuity," was also binary with values set equal to 1 when a respondent indicated that they would never take an annuity. Both variables were tested, since their positive outcome represented different choices.

The explanatory variables used in both models were as before, and included respondents' region of residence, age, gender, marital status, home ownership, children, work status, income, and education.[10] The results of logistic regressions are also described in Clark and Knox-Hayes (2007). Interestingly, none

---

[10] Notice, however, that the binary nature of both dependent variables violated the assumptions of OLS linear regression. Therefore logistic regression, which acts as a generalized linear model with binomially distributed errors, was used to estimate both sets of models. Since the β parameter estimates of a logistic regression represent log odds, which are difficult to interpret, the β parameters were converted into odds ratios.

of the socio-demographic variables was significant in either set of models. However, geographic regions were highly significant. Respondents from the northern and western regions of the United Kingdom—the North of England, Wales, and Scotland—were significantly more likely never to take an annuity. Individuals from the North of England were found to be also particularly unlikely always to take an annuity. Since socio-demographic variables were not significant, our findings may be an indication of a significant cultural effect—the existence of regional subcultures of risk propensity (as in Leyshon et al.'s 2004 "ecologies of finance"). The results were not representative of a lack of experience with private pensions, since 78 percent of respondents, and in particular 80 percent of those in the North, who would choose never to take an annuity held at least one supplementary pension entitlement. The results are consistent for those who would choose always to take an annuity, with over 80 percent of respondents holding at least one private pension entitlement.

We also sought to determine whether the "ecologies of finance" could be discerned in respondents who switched their intended behavior from would not to would, and from would to would not purchase an annuity. The same types of statistical tests were used, with the first dependent variable being coded in a binary fashion: 1 for switching between would to would not (285 respondents) and 0 for all other switching possibilities. The second dependent variable was likewise coded in a binary fashion: 1 for switching between would not to would (154 respondents) and 0 for all other switching possibilities. Here, it was shown that region of residence was again a significant determinant of the observed variance in responses. Moreover, the statistical results were symmetrical in that respondents in the South and Wales would be more likely to switch from would to would not purchase an annuity, just as respondents in the South and North would be less likely to switch from would not to would purchase an annuity. Importantly, however, it was also found that higher education is a significant correlate of the switch from would to would not purchase an annuity.

Granted, there are small numbers of respondents involved in intended switching behavior and the regions of residence are large and open to dispute. But the combination of education and region of residence suggests that Southern middle-class consumers of annuity products are rather sophisticated risk managers, in that their intended behavior accords, in part, with the expectations of those who advocate the use of annuities to protect against the risks of outliving resources: they would purchase an annuity when retiring at 60 and 65 years of age, thereby protecting their long-term retirement income. But they would switch out of annuity (if they could) at 70 and 75 years of age, either backing their ability to manage their own welfare or ensuring that they were able to leave a bequest. Notice that any switch in behavior is correlated with gender, with men being more willing than women to make such a decision, a result consistent with social science research on the confidence and risk tolerance in regard to pensions of men over women (see also Chapter 3).

These findings match those of Bao et al. (2003) and Henrich et al. (2005), who suggest that risk aversion among otherwise similar groups may vary according to "local" circumstances or culture. Our findings are also in line with Bertrand et al. (2005), who found in South Africa that the "local context" of risk-related decision-making can be significant in differentiating behavior within domestic financial and consumer markets. Likewise, unpublished research on the German market for retirement products suggests that there are significant regional differences in risk-taking, the purchase of equity-based investment policies, and confidence or otherwise in state-based social security. Closer to home, Nesbitt and Neary (2001) have shown that members of some South Asian groups located in London have tended to avoid private pensions, suggesting that culture, the choice of retirement plans, and decision-making can be tightly entwined at the subregional level.

There are, of course, other plausible interpretations of our findings that are less cultural and more related to economic imperatives (MacKenzie, 2006).[11] For example, could it be that many Northern respondents are averse to taking an annuity because of pessimistic assessments of their likely longevity? Many of the baby-boom generation underestimate their longevity; in this respect, Northern respondents may believe that their own circumstances are such that they will die before receiving the full benefits of an annuity and/or may be committed to passing on what limited savings they have in the form of a bequest.[12] Alternatively, they may believe that a lump sum plus state social security will be sufficient for their needs, particularly if they own their own home and expect that the cost of living in the North will be much less than living elsewhere (the South). "Getting by" may be the operative solution to retirement planning, rather than optimizing long-term income replacement. Should these expectations not be realized, one possible consequence may be that many Northern respondents will find their retirement savings quite inadequate over the long term, with significant consequences for regional economic welfare.

---

[11] For example, following from the work of Leyshon et al. (2004), Peter Sunley suggested that our results may reflect the advertising and marketing programs of companies that seek to sell annuities, presumably differentiating between Southern and Northern markets, levels of education, and perhaps occupation. We are not able to comment on this hypothesis or whether, in fact, vendors follow markets or create markets.

[12] Little is known about how people estimate their longevity. Anecdotal evidence suggests that many base estimates on their immediate forebears, especially parents and grandparents. However, it has been recognized that advances in public health over the post-1950 period have been such that the life-time of those born before 1914 has little relevance to the life-time of those born before 1939, just as their life-times have less relevance than expected for the expected life-times of the baby boom generation. See Fries (1980) for a prescient assessment of this point.

Implications and Conclusions

In this chapter, we posed questions concerning respondents' pension behavior in the context of risk and uncertainty. Those questions were framed by reference to the downturn in global financial markets in the aftermath of the TMT bubble, justified by an argument to the effect that human decision-making occurs at the intersection between cognitive capacity and the environment (time and space) in which decisions are taken. This argument was developed in the opening chapter of the book and can be found in Kahneman and Tversky (1979), Simon (1956), and Stanovich and West (2000). Specifically, we sought to provide a context-specific assessment of individual behavior, testing for correlations between individual responses and their socio-demographic characteristics and where they live. At the heart of the analysis was an assessment of the relative significance of geography and society in observed pension-related risk attitudes and behavior.

Two tests were deployed to evaluate the relationship between the number of supplementary pensions held by individuals, their region of residence, and their socio-demographic characteristics. Similarly, two tests were deployed to evaluate the relationship between respondents' willingness to take an annuity and their regional and socio-demographic characteristics. We found that women and people who care for children were significantly less likely to hold more than one pension. This finding is likely to be indicative of the shorter time that women spend working due to domestic care-giving, but may also indicate that women are less likely to switch jobs given their responsibilities. Not surprisingly, respondents who are older, have a higher level of education and household income, are single, and are home owners were found to be more likely to hold a larger number of private pensions. This finding is consistent with the probable greater earning capacity of those individuals. As expected, respondents who are younger, employed part-time, retired, or unemployed were found to be less likely to hold private pensions.

These results are consistent with our previous findings, to the effect that respondents' socio-demographic status including age, gender, income, marital status, and household characteristics are significant in differentiating financial decision-making. These results reinforce the findings of other scholars regarding the importance of financial literacy in the United Kingdom and elsewhere, especially considering the increasing responsibilities of individuals and their families for planning their future welfare (see, e.g., Lin and Lee, 2004; Worthington, 2006). There can be no doubt as to the sensitivity of UK social welfare to the individual and collective decision-making capacities of residents—this point has been made by many commentators over the past few decades and has been reinforced by the Turner Report on the current pensions crisis and its possible resolution through the NPSS (DWP, 2006). Even if state-provided social security benefits are indexed to average wage growth in the future, it will remain a modest component of household retirement income.

While socio-demographic variables explain considerable variance in the number of pensions held, this explanatory logic does not extend to their willingness to take or not take an annuity. Geography is more important in this respect. We found that individuals from the North of England, Wales, and Scotland were significantly more likely not to take up an annuity. Likewise, we find region of residence as well as education and gender significant correlates of systematic intentions to switch from an annuity to a lump sum by age. These results suggest the existence of significant "local" risk cultures driving individual considerations of retirement income risks.

This chapter challenges accepted notions of individual decision-making on two counts: by the status we attribute to "the environment" (an essential but under-studied element in financial decision-making) and by our empirical analysis, wherein it was demonstrated that there are systematic social and geographical effects on respondent risk management behavior—a finding consistent with integrated models of social behavior and cognition (see Krueger and Funder, 2004). Most importantly, our results suggest the existence of a differentiated UK "map" of individual risk predilections and behavior that has its roots (in part) in the social stratification of the UK population and (in part) in the coexistence of distinctive regional risk cultures (as suggested by Leyshon et al., 2004, 2006). To the extent that national risk cultures are believed consistent with universal models of rationality given the discounting of history and geography on the way toward a putative one-world, regional or intra-national risk cultures in the face of economic and social integration challenge assumptions of behavioral homogeneity.

Why is respondent financial decision-making affected (in part) by social status and (in part) by residential location? At this stage, we can only speculate. First, it should be recognized that human beings hardly ever "make decisions" if what is meant is a deliberate weighing-up of the costs and benefits of alternative options before choosing and taking action. Rather, as Kahneman (2003) suggested, most "decisions" are virtually automatic responses to changing circumstances guided by intuition, habit, or imitation (Chapter 2). In fact, some analysts argue that intuition is more often than not the "best" response to circumstances where the complexity of social life overwhelms simple rules or heuristics that have worked in the past (Gigerenzer et al., 1999). Why does it matter, then, if a person is young or old, educated or not, male or female, or rich or poor? One answer is that people do not *only* rely upon intuition; their heuristics imitate others whom they either admire or believe to represent their aspirations. To the extent that the media play a role in defining such role models, heuristics could be shared but highly differentiated by social class and aspiration (Clark et al., 2004).

This argument is a useful way of explaining the significance of social status. But it is somewhat removed from the nitty-gritty of where and how people live (Leyshon et al., 2004, 2006). Here, we would argue that "local" risk cultures could have a material significance when local circumstances are profoundly at odds with shared aspirations. For example, imagine that the nation-state is divided into a set of regions where one is extremely prosperous and the other extremely poor (ONS, 2006a, 2006b). In the former, growing employment opportunities ratchet up individual and household income to such an extent that the housing market follows suit. In the latter, declining employment opportunities and low rates of income growth are combined with a barely moving housing market. In the former, the labor market, the housing market, and even the financial market reinforce respondents' expectations of prosperity, whereas in the latter market capitalism only brings grief. Why should they (the latter) trust financial markets? In fact, why should they be anything other than dependent on existing state welfare and focused upon short-term benefit? By this logic, material circumstances reinforce (at the extreme) decision heuristics that become increasingly at odds with national norms and customs (a phenomenon widely recognized in psychology).[13]

We are not able to support this argument directly with empirical evidence. Indeed, to test the argument would require estimating individuals' discount functions against some kind of social reference discount function so as to measure divergence and the effects of local conditions and expectations on common expectations. Considering the paucity of empirical research on the nature and shape of individual discount functions, this is a research agenda of truly breathtaking scope and significance (noted by Ainslie, 2001; Laibson, 2003). Nonetheless, to assume that discount functions and decision frameworks are the same across regions and time within the United Kingdom—assumptions underpinning the government's NEST—may lead to long-term problems in the local performance of national pension institutions that could accentuate apparent income inequalities between regions.

---

[13] A point made by many social psychologists arguing that behavior is reinforced, positively and negatively, by the nature, frequency, and consistency of rewards. See Rachlin (2000) for the seminal exposition of the theory of reinforcement. Notice that the argument made in our paper has a materialist base; reinforcement can also have a nonmaterialist origin in the shared values of a community.

APPENDIX
Variable Coding Structure

Variable name	Variable description	Category	Associated value
Dependent variables			
Pension number	Number of private pensions held	0	0
		1	1
		2	2
		3	3
		4+	4
Always take an annuity	Response to four opportunities to accept an annuity at age 60, 65, 70, 75	Always yes (yyyy)	1
		Sequences of yes no maybe (ynmn)	0
Never take an annuity	Response to four opportunities to accept and annuity at age 60, 65, 70, 75	Always no (yyyy)	1
		Sequences yes no maybe (ynmm)	0
Independent variables			
Geographic region	Region of residence in the United Kingdom	London	1
		South	2
		Wales	3
		South West	4
		North	5
		Scotland	6
Female	Gender status	Male	0
		Female	1
Marital	Marital status	Married	1
		Divorced	2
		Single	3
Home owner	Ownership of residence	Home owner	1
		Non-home owner	0
Children	Child status	Children in household	1
		No children in household	0
Work	Employment status	Working	1
		Working part-time	2
		Retired	3
		Otherwise unemployed	4

Income	Annual household income	Less than £20,000 (low)	1
		£20,000–39,999 (middle)	2
		£40,000 or more (high)	3
Education	Level of educational qualification	No qualifications	1
		Intermediary/ technical	2
		University/ Professional	3

8
The "New" Paternalism

Many of our core economic and social institutions presume individual sovereignty—whether by design or by default, the individual is deemed the "responsible" agent of his or her future welfare. In private pension systems, this responsibility has grown in leaps and bounds as employers have jettisoned conventional defined benefit (DB) plans in favor of defined contribution (DC) plans or individual pension contracts. As we have seen, this type of saving scheme makes significant demands on individual reasoning and decision-making: it appears that few people are the sophisticated planners that theory assumes, and even fewer appear to be sophisticated market players willing and able to save for the future in the context of market volatility. In the United Kingdom, of course, the rapid decline of private DB pensions and the rather slow uptake of DC and individual pensions has brought forth an unexpected (for the government) pensions crisis (Clark, 2006a). Solutions likely to produce a reasonable level of income replacement upon retirement are hard to find.

In the United States, the costs and benefits of DC pension schemes for retirement welfare have been subject to searching scrutiny (see Ghilarducci, 2008; Munnell and Sundén, 2004). With the passage of the 2006 Pension Protection Act, Congress sought to reinforce tentative steps taken by plan sponsors and the financial services industry to assume greater responsibility for framing the risks assumed by plan participants and the realized value of accumulated pension savings. Benartzi and Thaler's Save More Tomorrow (SMT) regime (2005) has been widely touted as an effective mechanism for increasing the rate of contributions. Similarly, target-date funds offered by investment banks and mutual funds may provide workers with a once-and-for-all retirement plan that automatically adjusts their asset mix in relation to the years to retirement. The goal here is to reduce the risks associated with investment as participants near retirement. This option has attracted attention, in part, because the US Department of Labor deemed such funds an "appropriate" default option for employers who automatically enroll employees in 401(k) plans.

Auto-enrollment, opt-out rather than opt-in, contribution escalators, and prescreened investment options matched against participants' age and risk tolerance are elements of Thaler and Sunstein's libertarian paternalism (2003: 179): "an approach that preserves freedom of choice but that authorizes both private and public institutions to steer people in directions that will promote their welfare." Thaler and Sunstein (2008) have also suggested that the proper purpose of public and private institutions is to "nudge" people to act in ways consistent with their interests, recognizing that cognitive biases and anomalies tend to derail their best intentions. This stands in contrast to directing or requiring certain behavior; it is also rather different to mandatory pension savings schemes like those in Australia, New Zealand, and Sweden and rejected by the UK Pensions Commission (2005) when recommending the establishment of the National Pension Savings Scheme (NPSS). Thaler and Sunstein suggest what planners—"anyone who must design plans for others"—ought to do, given the welfare costs of ineffective decision-making.

Critics of finance-led neoliberalism argue that public and private institutions have shifted the risks associated with pension planning to people inadequately equipped to handle the issues (Langley, 2008). Missing here is recognition of the evolving agenda that goes under the banner of the "new" paternalism. But missing from both research programs is an appreciation of the nature and scope of individuals' expressed preferences (if any) in being involved (or not) in the retirement planning process. This topic is explored here with reference to the degree to which a representative sample of UK residents would prefer to be consulted in plan-initiated asset-switching prior to respondents' retirement. As indicated, this is a significant issue for the design of retirement investment products and bears on the mandate claimed by the "new" paternalism. Following on from our previous chapters on pension planning, we estimate the correlates of consultation over asset allocation leading up to retirement.

Before summarizing the results of our statistical analysis (reported in detail in Clark and Knox-Hayes, 2009), we consider the design of retirement funds, given the growing number of UK DC and personal pensions. This is followed by an analytical framework for explicating the nature and scope of paternalism. Beginning with Thaler and Sunstein (2003), we suggest a framework that can allow a more nuanced understanding of the scope of paternalism. It is shown that those respondents who would expect consultation over the transition to retirement are those whom we found previously to be committed to pension planning: older, higher-income people who appreciate the significance of supplementary pension plans for their retirement income (Chapters

3 and 4). As well, we found a distinctive geographical pattern in responses, partially validating our finding in Chapter 6 about the existence of "regional ecologies" of finance. Importantly, there was a group of respondents who could not venture an opinion. These respondents are best characterized as discouraged or disaffected.

Our approach to these issues is to mix theoretical and practical issues related to behavior and the design of retirement products with our national database of respondent attitudes, opinions, and behavioral predispositions. It is gratifying to find significantly the same types of socioeconomic correlates as before. In terms of the book overall, this chapter makes the link between previous findings and the argument made by advocates of the "new" paternalism that there are ready solutions to hand. As is shown, we are not convinced. As such, this chapter is a reference point for the final chapter of the book, which focuses on institutional design and policy.

Risk and Uncertainty

The research program initiated by Simon (1956) and Kahneman and Tversky (1979) assumed substantive rationality but focused on cognitive performance, arguing that humans are subject to a variety of systematic biases and shortcomings. For example, to the extent that people have coherent discount functions (many do not), they are preoccupied with the immediate future and find it difficult to carry through on plans. These "traits" have been shown to be widely shared, though there remains debate over the degree to which conscious deliberation and education can overcome or in some sense compensate for the negative effects of these shortcomings. We noted in Chapter 4 that recent research on consciousness and the link between intention, deliberation, and action raises profound questions about the degree to which people do anything more than respond to sensory impulses—at issue is the status of individual preferences and volition (see Rachlin, 2000). We have argued that recent policy innovations that go under the banner of the "new" paternalism appear to assume that many people (except sensible "planners") are on automatic pilot. Hence, one way to encourage higher levels of pension saving is to introduce contribution escalators that rely upon automatic increments rather than rely upon individuals for a deliberate reallocation of consumption for saving now and in the future (Benartzi and Thaler, 2005).

The significance of these traits can also be seen if we refer to the aptitudes and skills required for effective performance in global financial markets. Without suggesting that our list is definitive, our assessment of the competence of financial decision-makers identified the following desirable attributes:

being comfortable in the probability domain, being numerate and able to calculate the future value of alternative courses of action, being able to distinguish between underlying patterns and contingent events, being able to process information in an efficient and timely manner, and being able to cut through the clutter of events and apply the relevant decision rules and techniques (Clark et al., 2007). As the global credit crisis has demonstrated, financial markets can be extremely demanding environments. Being able to judge the significance of market movements time and again without hesitation and with a complement of skills equal to at least professional market players is very important (as evident in Chapter 5).

Translating these observations about financial competence into the realm of self-directed pension savings and investment, it is widely accepted that self-interest and an everyday knowledge and understanding of financial markets are inadequate, given the dependence of so many people on supplementary savings and investments for long-term welfare (Ghilarducci, 2008). To illustrate, leaving people to make their own decisions about asset allocation as they approach retirement may mean that they carry too much risk (e.g., an equities exposure) for too long, thereby becoming vulnerable to the state of financial markets at the point where they must switch their accumulated assets into some form of annuity (Poterba, 2006). Here, at least three types of behavioral biases or shortcomings may conspire to discount significantly future welfare. Inertia is one issue (Samuelson and Zeckhauser, 1988), lack of attention is another consideration (Gabaix et al., 2006), and under- or overreaction to recent stock market events may be another issue (see Shiller, 2000).

For those knowledgeable about life-cycle risk and asset allocation, a plan participant's age and years to retirement as well as his or her risk tolerance are crucial variables in constructing "optimal" investment portfolios. To illustrate, Figure 8.1 reproduces a diagrammatic representation of Barclays (now Blackstone) Global Investors (BGI) LifePath Portfolio (a successful DC product in the UK market). This is typical of target-date funds, in that in the early years the fund asset allocation is such that growth-with-risk is the operative strategy, whereas in later years conservation of value of accumulated capital through asset diversification becomes the operative strategy. In this case, identified asset classes are entirely conventional and would be recognized as such by any participant who has a modicum of interest in the current offerings of investment companies.

Whether the asset-mix adjustment process ought to be continuous or staged, whether it ought to have a shallow or steep glide-path approaching retirement, and whether plan participants should be required to be actively involved in each stage remain to be resolved. As well, there remain reasonable concerns about the costs of such portfolios and whether those who adopt such retirement products must bear relatively high costs and market risks in the first

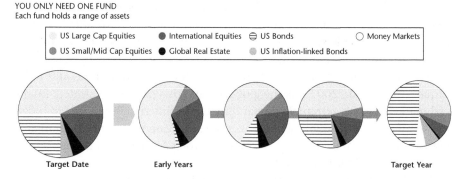

Figure 8.1. Blackstone's target date fund
Source: http://www.bgidcsolutions.com/lifepath/how-lifepath-works.html (July 2008).

stages of the process. Presumably, the chosen investment funds that make up the BGI portfolio come entirely from Blackstone. In such cases, there may be reasonable concerns over the relative costs and performance of "captive" participants.

Paternalism and Coercion

For plan sponsors, target-date funds appear to resolve a number of concerns about participants' lack of attention to their long-term welfare. Moving from opt-in to opt-out, auto-enrollment may be thought to take advantage of participants' inertia. Combined with SMT automatic contribution escalators, a much higher proportion of eligible workers are likely to save more for retirement than otherwise the case. With the automatic enrollment of participants into target-date funds, they are also likely to be more effective in risk management according to age and the expected date of retirement. In a sense, workers may be "made" into effective market participants and their behavior "channeled" in accordance with the best available evidence on cognition, behavior, and commitment over time (see Preda, 2004). Thaler and Sunstein (2003: 175) termed this bundle of policies "libertarian paternalism," in that plan designers seek to influence "the choices of the affected parties in a way that will make those parties better off."

For some, paternalism is a proper concern for the welfare of children (Nozick, 1997). For others, it is aligned with tradition—legitimated by a declared concern for the welfare of society. And for yet others, paternalism is shorthand for hierarchical societies based upon privilege and control over the mass of people's well-being and prospects (as suggested in Atiyah's 1979 account of the transition from feudalism and the rise of "freedom of contract"). Forgotten, perhaps, in the debate in the pensions literature over paternalism is the fact that the spread of DB pensions throughout Western economies over the twentieth century can be interpreted as an expression of benevolent paternalism. For union and management executives, collectively negotiated wage-based deductions for retirement served many purposes, including increasing the rate of savings among hourly workers who would not have otherwise made such long-term commitments (Clark, 2000). Mandatory NPSSs are based on much the same motives.

Recognizing the various ways in which paternalism can be defined (see Dworkin, 1972; Lively, 1979; Shiffrin, 2000), for the purposes of this chapter we define paternalism in the following way: assuming B (the object of paternalism) is a mature person legally and morally responsible for their actions, A trumps B's exercise of free will and judgment and, as a result, the outcome of A's actions (or non-actions) on behalf of B is to the benefit of B (and may or may not be to the benefit of A or his or her proxies). To trump another's exercise of free will and judgment must involve more than a calculation of net benefit—on this basis, it would be too easy for A (an institution or person in authority) to legitimate arbitrary intervention. It must be based upon a justified claim of superior expertise, knowledge, or commitment to social welfare that transcends the circumstances of any individual. To avoid the charge of coercion, it must also be accompanied ex ante or ex post by B's consent (explicit or implicit). This is, clearly, a complex web of statements and conditions and properly so given the liberal democratic presumption in favor of individual autonomy (Holmes, 1993).

Notice that this definition is built on a set of background principles: to trump B's exercise of free will and judgment is to put aside his or her rights (not deny his or her rights) on a particular issue; to promote the net benefit of B is essential, just as any benefit to A (if any) should be incidental; benefits to A should be justified on some "reasonable" grounds consistent with B's interests; consistency in the treatment of others is essential if paternalism is to avoid the charge of being arbitrary; and A should be subject to the consent of B if paternalism is to avoid being tarred with the brush of coercion. We can see, then, why the opt-out option in Thaler and Sunstein's scheme is so important. Still, the design issue is more problematic than we might have imagined—for example, in designing DC and accumulation accounts based on the principles of the "new" paternalism, it may be difficult to prove ex ante or ex post that such funds are to the net benefit of the recipient. The former requires complex simulations which may be inconclusive, whereas the latter may be so far into the future and subject to such uncertainty that demonstrating net benefit where time cannot be reversed is impossible.

For all the virtues of paternalism, some nonetheless argue that paternalism is akin to coercion (and is consequently illegitimate). In order to distinguish between the two, we define coercion in the following manner (see Wertheimer, 1987): assuming B is a mature person legally and morally responsible for his or her actions, A subrogates B's exercise of free will and judgment and, as a result, A and B's actions (or non-actions) are not to the benefit of B (and may directly benefit A or his or her proxies). Domination of B can take a variety of forms including force, threats, and the lack of any other viable option (including noncompliance). In all cases, instruments of coercion must be perceived by B as credible and to have consequences greater than the immediate costs of compliance. Coercion may be personal, and may also be aimed at a class or group of people who have neither the capacity to resist nor a reasonable claim on another institution or authority for protection. Again, this is a complex web of statements and conditions. By our account, coercion is more than an issue of costs and benefits. It also involves harm done to others by reason of the denial of their free will and judgment.

Is the "new" paternalism coercive? While it may be difficult to demonstrate the net benefit to B of acting on his or her behalf, except in an abstract manner, assuming A (a company or plan sponsor) abides by the principles and practices of fiduciary duty, auto-enrollment and the package of related policies should not benefit A (except, perhaps, in an incidental manner). Assuming auto-enrollment is not accompanied by the use of force, threats, or intimidation, the "new" paternalism is not directly coercive. But this does depend upon the formal process of auto-enrollment and the ease of executing the opt-out option (if desired). Auto-enrollment accompanied by long periods of "lock-in" and then lack of notification to participants of the timing and means by which the opt-out option is exercised may effectively coerce continuing enrollment albeit by neglect or default. Where a heavy burden is placed on the participant in exercising opt-out options, well beyond the immediate costs of remaining with the plan, the "new" paternalism may be coercive (albeit lacking force or threats).

In the consumer world, there are many examples of unscrupulous companies taking advantage of the insights gleaned from the behavioral revolution. The automatic rollover of monthly and yearly direct debts, the obscure placement of opt-out tick-boxes in complex enrollment forms, and the onus placed on customers to apply for so-called "automatic" discount entitlements are all designed to take advantage of customers' lack of attention. Inevitably, most financial issues are complex and difficult to understand in terms of their contracts and long-term consequences. This may have implications for certain classes of people (see Stanovich and West, 2000).

Consultation and Tacit Consent

At the heart of the "new" paternalism is an asymmetrical relationship between "planners" and the intended beneficiaries of plans. This relationship can be described as follows: being aware of the finance-related cognitive shortcomings of beneficiaries, planners design and put in place decision frameworks that channel and prompt behavior consistent with beneficiaries' long-term interests. Either unaware of, or inattentive to, their cognitive shortcomings, many beneficiaries implicitly consent to participate in plans; they do so either by signing an employment contract, or by trusting in the best intentions of planners, or both. One way or another, planners have the knowledge and expertise to anticipate the likely actions of beneficiaries. In this sense, planners control the terms and conditions of plans and hence the likely behavior of plan beneficiaries. This is clearly paternalistic (in a good sense).

However, this is a privileged position. And it carries certain obligations, one of which is not to exploit knowledge of beneficiaries' shortcomings. In this regard, it is possible that planners are subject to legally inscribed duties to beneficiaries and may be sanctioned for any failure to fulfill those duties. Of course, one of the problems in this relationship is its asymmetrical character—it may be very difficult for any beneficiary to obtain the knowledge and information necessary to challenge the actions of the planner (Clark, 2006b). More generally, it could be argued that the "power" of the planner is such that he or she is obliged to test the consent of beneficiaries to these arrangements, especially when circumstances change on either or both sides of the relationship. In any event, given the inevitable problems in demonstrating the long-term benefits of paternalism, the planner may come to see that there is a relationship between testing for consent and maintaining the legitimacy of the arrangements.

Surely, the virtue of the "new" paternalism's automatic enrollment process is the opt-out option. Doesn't the fact that the vast majority of participants choose NOT to exercise this option warrant the related claim by planners that those participants tacitly consent to the arrangement? The status of tacit or implicit consent is more problematic than perhaps recognized. Simmons (1979: 79–80) used an example from the management of a meeting to suggest that, as normally understood, tacit consent can be claimed only (*a*) when participants understand the significance of their lack of objection; (*b*) when participants are aware of the procedures that may be used to voice objection; and (*c*) when participants appreciate the need for timely objection consistent with the implication of the related policies. Where these conditions are not met by happenstance, the availability of an opt-out option would not be sufficient to claim participants' implied consent. Where these conditions were not met by reason of planners' deliberate obfuscation of the terms and conditions for agreement, it could be argued that consent has been coerced and is illegitimate.

Simmons (1979: 81–2) goes on to introduce two further conditions (reflecting comments made above) for the claim of consent to be legitimate: (*d*) the mechanisms for voicing disagreement must be reasonably exercised, and (*e*) the costs of voicing disagreement must not be punitive (over and above the costs of alternative courses of action). Both are entirely appropriate and, with the other three conditions, provide a template by which to judge the efficacy of an opt-out option. These issues are, as well, a checklist easily incorporated into a plan's trust deed and government rules and regulations. An independent auditor could be charged with the responsibility of certifying their utility and reasonableness (subject to a code of best practice or regulation). But this may not be sufficient. The asymmetrical problem is that tacit consent is claimed by planners, not "given" by beneficiaries. So, a further test of consent may be necessary, one that requires actions by beneficiaries that indicate *consensual* participation in schemes and their conditions. Given the long time horizon over which target-date funds function and the sensitivity of beneficiaries to its performance, it may be appropriate to test participants "consent" by simple checks on their attention to the asset switching process.

Notwithstanding the published research devoted to behavioral anomalies in DC and self-directed pension plans, there are few papers that have sought to test participants as regards the nature and frequency of desired consultation. If there are surveys on these issues, the results remain with the consulting companies. For this chapter, we return once again to the representative sample survey of UK participants in DC and self-directed plans sponsored by Mercer Human Resource Consulting (see Chapter 3). Included in the nearly eighty questions posed to respondents in intensive telephone interview was a question about the extent to which respondents would wish to be consulted in plan-initiated asset switching as participants near the date of retirement. The question (#48) was posed in the following manner: "*As people approach retirement, it may make sense for them to move their pension savings out of high-risk investments such as equities, so that they do not suffer from a short-term fall in the stock market. Which of the following would you prefer to happen in the years leading up to your retirement?*"

A set of five options were provided to respondents, ranging from a high level of individual control (Option 1) through to a low level of individual control (Option 3), don't know (Option 4), and n/a (Option 5). Specifically, Option 1 provided "*For your pension provider to advise you to move your savings to lower risk funds, but leave it to you to decide what (if any) changes to make.*" Option 2 provided "*For your pension provider to advise you to move your savings to lower risk funds, and to recommend a selection of funds to you.*" Option 3 provided "*For your pension provider to automatically move your savings to lower risk funds and to tell you that they are doing this and why.*" The survey did not directly test the expressed commitment or otherwise of respondents to target-date funds, but it did test respondents on their willingness or otherwise to let the plan sponsor take responsibility for a crucial element in target-date funds: the asset-switching process as the participant nears retirement.

Many social scientists are skeptical of expressed preferences, preferring observed actions as opposed to the views of respondents who may seek interviewer approval. Nonetheless, there is evidence from a variety of disciplines that, in domain-specific tests of opinion, there is a close relationship between expressed preferences and action (see Dohmen et al., 2005; Dorn and Huberman, 2005; Weber et al., 2002). In any event, whether the "new" paternalism is effective in setting the parameters for savings plans that rely upon (for example) auto-enrollment and target-date funds may depend upon meeting the range of expectations among plan participants. Otherwise, there is a possibility of taking for granted "consent" and unknowingly "coercing" choice when in fact participants would wish to be directly consulted and involved.

Statistical Strategy and Results

The source of the survey data and the methods used to collect responses consistent with the interests of the survey sponsor are described in previous chapters. It is important to emphasize at this juncture that the survey was intended to be a representative sample of the UK population enrolled in DC and related self-directed pension plans. Those *only* enrolled in DB plans were excluded. Likewise, it was intended that respondents would come from the private sector, although this condition had to be relaxed as it became difficult to realize the planned number of respondents. Respondents were drawn for telephone interview from lists of potential respondents—to be included in the survey, they had to be currently employed in an organization with more than ten employees. Included in the data collection were a number of socioeconomic and demographic characteristics including gender, age, income, occupation, and spousal circumstances. We have shown that spousal circumstance is a significant correlate of risk preferences and planning for retirement.

For the purposes of this study, we began by tabulating the responses to Question 48 by the five options (including n/a) by income. These are summarized in Table 8.1. A number of observations can be made that bear upon subsequent analysis and interpretation of the statistical results. The n/a group was very small, and was not included in the analysis of the correlates of

Saving for Retirement

Table 8.1. Level of consultation desired by income

	£15,000 or less	£15,000–£25,000	£25,001–£40,000	£40,001–£65,000	£65,001 or more	Total
High consultation	25	41	42	22	3	133
Some consultation	98	168	138	69	20	493
No consultation	26	38	38	11	3	116
Don't know	42	50	27	13	1	133
N/A	4	3	7	1	1	16
Total	195	300	252	116	28	891

consultation. The Option 4 "Don't Know" group was surprisingly large and was also excluded from the analysis of the correlates of desired consultation. Notice how in the related survey (Chapters 5 and 6) of respondents drawn at the peak of the bubble from a large multinational financial services company operating in London fewer than 8 percent of respondents checked the "Don't Know" option and about 10 percent skipped the question (probably registering n/a). Finally, in terms of the distribution of respondents by income classes, the fact that most respondents were clustered at the lower end of the distribution reflects the strategy of the survey designers and our interest in a representative UK sample. We return to the significance of income for desired consultation in the following subsections.

Correlates of an expressed preference for consultation

To improve our understanding of the correlates of an expressed preference for consultation, we constructed a dependent variable out of the first three options such that it can be interpreted as the likelihood that respondents expressed a strong preference for consultation (i.e., Option 1). Drawing upon the previous chapters on the correlates of retirement planning and the correlates of risk preference in asset allocation, we sought to determine whether age, gender, income, the expected replacement rate, and household circumstances were significant correlates or predictors of a strong preference for consultation. Recall how in previous chapters we demonstrated that the older the respondents, the higher their income; the fact that they recognized that supplementary pensions are designed to replace earned income; and the fact that some households included spouses with a similar entitlement were all positive predictors of an increasing likelihood that respondents believed retirement planning to be important. We had every expectation that these variables would be also important in the demand for consultation in sponsor-prompted asset-switching leading up to retirement.

The "New" Paternalism

In Clark and Knox-Hayes (2009), we report in detail the econometric results based on the socioeconomic variables and a number of other risk-related variables. In this case, based on the sample excluding the "Don't Knows" and n/a respondents, it was found that being female and having a higher income were significant and positive predictors of an expected high level of consultation. By contrast, having a spouse with a related entitlement, relying upon another for retirement welfare, and being older were significant but negative, suggesting that these types of respondents were less likely to demand a high level of consultation. Notice, as well, that region of residence was a significant and positive predictor of an expressed preference for a high level of consultation *if* respondents lived in either East Anglia or Scotland (compared to London). As we have noted in Chapter 7, there is a distinctive geography to UK private pension provision and expectations that deserves more scrutiny (see also Sunley, 2000).

Not significant was the expected value of the pension in terms of retirement income. We also sought to determine whether respondent confidence in retirement planning or their risk preferences and tolerance were significant variables affecting the demand for consultation. In the survey, a number of questions were asked regarding respondents' confidence in making "financial preparations" for retirement (Q11), their understanding of the "savings and investment options available for long-term financial planning" (Q12), and their confidence in the accumulation of money for retirement (Q44b). We asked respondents for their level of satisfaction with the company pension plan (Q20b, Q20c) and whether they would like to plan for retirement but don't know what to do (Q15) or can't afford to do so (Q16). On a yearly basis, we asked how often they initiate switching of assets between investments so as to reap "the best possible return" (Q47) and how risk-tolerant they were, providing a choice between "the best possible growth" in savings and "safe and secure savings" (Q25).

Having sifted these related questions for evidence of cross-correlation, it was found that those who expressed a high level of confidence in their understanding of savings and investment options for long-term planning and those who expressed a high degree of confidence in the value of the company plan were more likely to demand a high level of consultation over the transition to retirement. In a series of other tests, where we restricted the number of variables included in the analysis, and hence expanded the number of observations available for analysis, the only other variable found significant and with a positive sign was the realization that supplementary pensions replace a portion of earned income. Taken with the significant parameters on gender, higher income, and family circumstances, there seemed to be a group of active planners in our sample who were very conscious of the role that their plan will play in their future retirement welfare. Equally, there are respondents who, by virtue of their lack of confidence in themselves and others, were less likely to demand a high level of consultation.

Correlates of no expressed preference for consultation

We now turn to the group of respondents who were either unwilling or not able to express a preference for one of three levels of consultation. In this case, the whole sample was divided into two groups, "Know" (0) and "Don't Know" (1), and a logit model deployed to estimate the probability of "Don't Know" using the full range of variables identified above. Unfortunately, there were a large number of respondents who failed to indicate a response to each and every issue. Nonetheless, the estimated model reported in Clark and Knox-Hayes (2009) was robust.

It was apparent that some of the socioeconomic variables identified above and in previous chapters as significant predictors of the likelihood of retirement planning were not significant. Even so, consistent with our findings above, those respondents less likely to express "Don't Know" were those with higher income and those who recognized the value of the pension for future income, while those who indicated a reliance on others were more likely to indicate "Don't Know." Similarly, those who expressed a high level of confidence in their understanding of savings and investments and those who expressed a high level of confidence in the company savings plan were less likely to indicate "Don't Know." Those more likely to express "Don't Know" were those who were less likely to switch savings and investments in planning for the future and those who indicated that they could not afford to plan their retirement. Respondents in the "Don't Know" category seemed to see the consultation process as irrelevant to their interests, or they lacked confidence in the planning process.

In sum, we might have expected these results. Lack of confidence in pension planning and a lack of engagement in retirement savings plans are consistent with disengagement with sponsor-led management of the transition to retirement. That neither region of residence, nor age, nor gender, nor spousal pension entitlement was positively or negatively significant in determining the "Don't Know" category suggests that those respondents who fell into this category could have strong views about the perceived quality of plan governance and proffered advice and would act accordingly in terms of their willing engagement in the retirement saving process.

Diversity, Paternalism, and Policy

The results of our analysis are broadly consistent with those reported in previous chapters. Here, the range of significant socioeconomic variables was quite similar to that found for retirement planning. The significance of age, gender, income, spousal entitlement, and reliance on another provides a compelling set of socioeconomic indicators (positive and negative) as regards the types of respondents who may demand a high level of consultation in planning for retirement. These findings also reinforce related arguments in the literature about the rate of return on "attention" and whether plan participants appreciate the value of a supplementary pension for their long-term retirement income (see generally Gabaix et al., 2006).

In this section we aim to tease out implications from the statistical results so as to interrogate claims made on behalf of the "new" paternalism. By doing so, we emphasize a crucial finding embedded in the results summarized above: the apparent diversity of opinions and expectations of those respondents enrolled in DC and self-directed plans.

Our respondents were divided into two groups, one larger than the other: those who desire some level of consultation when plan sponsors initiate asset-switching prior to retirement; and those who really "Don't Know" what to do in these circumstances. The first, larger group would not be satisfied with an automatic asset-switching process; they would demand some level of consultation even if they would then accede to sponsor policy. We could break this group into three by the level of expected consultation. There are, as well, respondents who are in any event quite active in initiating asset-switching. They may be a subset of each, or a group in their own right. Whether these groups would appear in each and every plan is impossible to determine; the offer of three consultation options may have prompted respondents to make a response that could change as the nature and number of options change. We may have identified a similar effect argued to be characteristic of plan participants when offered asset classes to allocate their available funds (see Benartzi and Thaler, 2001). Still, that respondents would respond positively but differentially to consultation is an important finding.

What of the "Don't Know" group? On the face of it, this group is clearly a prime target for the policies of the "new" paternalism; lack of recognition of the significance of the issue is surely one condition for prompting enrollment on their behalf. But, our results suggest that such a response might cut against a number of rather different considerations behind the "Don't Know" response. In fact, there could be another three different types of respondents: those who do not appreciate the significance of pension plans for their retirement income and could be designated *ignorant*; those who simply can't afford to save for retirement and therefore don't see the relevance of the issue and could be designated *discouraged*; and those who do not have confidence in their employer and could be designated *disaffected*. Each of these subgroups may include a number who are just not engaged in the issue. It would be consistent with the motives behind the "new" paternalism to enroll the ignorant. But can the same be said for those discouraged and disaffected? The former may have other immediate goals like saving for the purchase of a house. The latter may have good reason to be doubtful about the integrity of the plan sponsor.

Auto-enrollment without an adequate process of consultation may be judged coercive by those who demand a high level of engagement or are, themselves, quite active in the asset allocation process. Equally, auto-enrollment of those discouraged and disaffected may be judged coercive, especially if the process of opt-out is cumbersome, difficult to activate, and demanding in terms of the required cognitive and emotional skills (*contra* Bodie and Treussard, 2007). In any event, the coexistence of a diversity of participants' expectations and (implied) motives if systematically replicated in UK DC and self-directed plans suggests that plan sponsors have a significant obligation to consult with participants regarding their needs and expectations. We would argue, as well, that DC plan sponsors may have to take seriously the design and governance of their plans in relation to participants' expectations—perhaps more so than is currently the case.

Implications and Conclusions

When considered over the past 150 years, Anglo-American pension policy has been preoccupied with individual sovereignty. Debate over old-age poverty in late nineteenth-century London combined a genuine concern for the welfare of older people with the liberalism of John Stuart Mill emphasizing the proper responsibility of individuals for their own well-being and that of their families. When debating the passage of the 1908 Old Age Pension, the government returned time and again to both the future cost of the state pension and the importance of balancing its paid benefit against the costs of disincentives on individuals to save for themselves (Clark and Whiteside, 2003). For all the significance of the Beveridge Report of 1944 for welfare and national health policy, the value of the British State Basic Pension remains modest compared to OECD countries (OECD, 2007). The recent Pensions Commission's recommendations (2005) are based on encouraging individual thrift notwithstanding a commitment to improving the welfare of the poorest people in retirement.

Debate over old-age poverty in the late nineteenth century was also preoccupied with the notion of moral worthiness: those people who by no reason other than the adverse circumstances of the time were deemed worthy of receiving state support. Equally, there were those by reason of their own fecklessness who were deemed not worthy of receiving state support. If once framed in terms of worthiness, contemporary market liberalism is not obviously about individual moral virtue. But it is committed to the principle of individual autonomy and encouraging a clear link between saving for tomorrow and future retirement income. So, it may seem surprising that the "new" paternalism is willing to trump individual decision-making with auto-enrollment, directed savings vehicles, and behavioral prompts that have as their goal real outcomes that are "better" than those which individuals may be able to achieve on their own account. This policy recipe would appear to be the antithesis of what many critics have claimed to be a pervasive neoliberal turn to the individualization of welfare (see Langley, 2008).

Whereas Victorians blamed feckless individual behavior and thereby absolved the state of responsibility, modern liberals have argued for greater financial literacy as a means of empowering individual responsibility for financial decision-making (absolving the state of responsibility for income inequalities resulting from apparent differential individual performance in financial markets). By contrast, the "new" paternalism has a radical interpretation of the behavioral revolution: given apparent cognitive biases and anomalies, it is wrong to "blame" people for their short-termism and it is a mistake to believe that people have the cognitive skills to function in financial markets to the level needed to be effective decision-makers. By this logic, financial literacy is simply a panacea rather than a solution to profound human traits. The "new" paternalism is hence motivated by a realization that exhortations to be more "moral" or more "rational" lack credibility in the face of the findings of behavioral research: hence, the lessons drawn for the design of pension institutions and policy (see Mitchell and Utkus, 2004*b*).

Nonetheless, the "new" paternalism remains deeply entwined with liberalism (as demonstrated by Thaler and Sunstein, 2003). This can be seen in a number of ways, most obviously in the obligations owed by pension "planners" to individual plan participants and beneficiaries. The presumption in favor of planners' expertise compared to plan participants must be balanced by the duties and obligations that planners owe participants. Otherwise, the "new" paternalism could become authoritarian and even exploitative of participants, given the limited mechanisms available for participants individually and collectively to affect the actions of planners. Whereas the adoption of DC and self-directed pension plans and the closure of DB plans by employers were intended to shift the responsibility and risks of pension and retirement planning to employees, the "new" paternalism has brought back to plan sponsors a heavy responsibility for ensuring the equitable governance and management of plans. If, as yet, not entirely recognized as such in the United Kingdom, the 2006 passage of the Pension Protection Act in the United States has set the agenda for greater regulation of such plans.

At another level, however, the "new" paternalism also runs the risk of denying the diversity of interests and expectations of plan participants by favoring frameworks that apply to all enrolled participants. There is little doubt that auto-enrollment, related investment products, and common decision cues effectively economize on the costs of pension provision for those covered. But, as we have sought to show in this chapter, there may be considerable differences in the degree of consultation on core elements of the "new" paternalism agenda expected by plan participants. These expectations have a legitimate claim to be canvassed and heard; to deny their veracity would be to deny the respect owed to citizens' expressions of interest and commitment (Pettit, 2007). Granted, not all participants may be so motivated; our results are clear on this count—higher income participants with a significant stake in the outcome of planners' decision frameworks are likely to claim center stage, while most other participants may be acquiescent.

Our results also suggest that there are participants who are neither engaged in saving for the future nor confident in the capacity of plan sponsors and their planners to make a difference to retirement prospects. Here, the "new" paternalism faces a significant challenge: to respond in effective ways to those seeking a certain level of engagement in the planning process while mobilizing the commitment of those otherwise discouraged and disenchanted and who may see auto-enrollment as simply an unwarranted tax on their incomes. These groups may be precisely those whom pension planners would seek to have enrolled in these types of programs. But they may be neither entirely willing nor entirely grateful for the experience. It is their interests that may have to be "trumped" by pension planners eager to affect the long-term welfare of all participants. In this case, the "new" paternalism could become the "new" authoritarianism.

9

Pension Adequacy and Sustainability

The new paternalism has been taken up by governments around the world. As Chapter 8 illustrated, however, its policy and practice contain certain fundamental contradictions. First, its advocates assume that individual choice and responsibility are almost always superior to collective risk sharing (especially under the auspices of the state) yet acknowledge that individuals often make less than competent decisions about issues relating to their well-being. Second, it is posited that the corrective to less than effective decision-making is not a transfer of risk away from the individual, but rather the construction of "choice architectures" that "nudge" pension plan members, purchasers of health insurance, etc. into making the "right" choices. In doing so, they acknowledge the importance of context but limit its reach. Third, by shifting the policy focus to individual-level decision-making, attention is deflected from the way in which choice is circumscribed. Individuals cannot trade autonomy for security; there is no choice to shift the risk back onto the firm or the state.

For these reasons, we are concerned about the implications of soft paternalism in pension policy; at the same time, however, we recognize that a "soft paternalist" approach can improve aspects of plan design, where the goal is to increase rates of enrollment and/or contributions. This tension goes to the heart of the approach taken in this book. On one hand, we are concerned with the "good enough" politics of addressing existing pension policy and institutions. On the other hand, we believe that taking seriously issues of cognitive capabilities, the context of decision-making, the role of emotions, and the distribution of social capital related to financial capability can be a way forward in promoting a "progressive" agenda of pension adequacy and sustainability.

In this chapter, we elaborate the implications of the preceding eight chapters for both "good enough" and "progressive" pensions policy. This involves looking at best practice in defined contribution (DC) plan design and governance, and the design of national pension institutions. Before turning to this question, however, it is worth revisiting the fundamental issues, touched on in the first chapter, of the role of occupational pensions and their contribution to the efficacy of a multipillar system. Within EU, *adequacy* and *sustainability* are the yardsticks by which pension systems are measured. Notice that choice and autonomy do not figure. How are adequacy and sustainability defined, and by whom? We would contend that there are two equally important aspects of sustainability: economic sustainability, or affordability; and social sustainability, or *equity*. In other words, a pension system is not sustainable in the long run if it is affordable (for the state and employers) but increases economic and social polarization. In some sense, this is also a core aspect of adequacy; however, we want to distinguish between a system that is defined as adequate because it prevents pensioners from falling into absolute poverty while increasing economic inequality, and one that is adequate in terms of providing a decent standard of living in old age for all.

This question of definitions is an issue for the United Kingdom. In the 2007 White Paper, the Government reiterated that the state has a responsibility to protect is citizens from insecurity and poverty in retirement, and that reforms to the UK pension system had to satisfy five criteria (DWP, 2006): promote personal responsibility, be fair (in particular, to women and caregivers), improve simplicity, be affordable, and offer a sustainable solution that commands a national consensus. Tony Blair, in the Prime Minister's foreword, also made reference to the contributory principle, further drawing a line under debates (which subsequently reemerged under the Conservative–Liberal Democrat coalition) about the conversion of the Basic State Pension (BSP) to a universal citizen's pension, as suggested by the Pensions Commission.

Notice that personal responsibility was the first criterion, with fairness second. Equality, or the reduction of inequality, did not make the list. Moreover, by equating unfairness with the disadvantages faced by women and caregivers, the White Paper implicitly defined fairness in the minimal sense of bringing those currently most disadvantaged by the system up to the level of median or norm (Strauss, 2009c). But what if the median, or norm, fails to protect pensioners from poverty and/or insecurity? The question of adequacy remains as the most important issue to be confronted and is considered again in the final section of this chapter.

Employer Commitment

In their heyday, defined benefit (DB) pensions were thought consistent with corporations' management of human resources. Unions also valued DB pensions because this type of benefit provided a means of rewarding older workers for loyalty and smoothing the transition to retirement on a modest

Pension Adequacy and Sustainability

supplementary income. There are two explanations for this alignment of interests: in terms of the acquisition of skill and labor productivity, it was assumed that older workers embodied vital on-the-job knowledge needed to exploit company-specific assets (Williamson, 1995). Similarly, it was believed that the skills of older workers become increasingly nontransferable with years of service in the employ of particular companies. One way or the other, the tacit knowledge and expertise held by older workers was deemed consistent with corporations' long-term profitability. In this sense, labor and managements' interests in the nature and value of provided retirement benefits were closely aligned (see Konzelmann et al., 2010). Several sectors of industry remain committed to DB pensions for these reasons.

However, it is apparent that many plan sponsors and younger workers do not value DB pensions, for various reasons. In the first place, it is apparent that many companies see DB pensions as a constraint on competitive strategy as well as a drain on financial resources (Monk, 2008). On the other side of the equation, younger workers do not value DB pensions as do older workers because younger workers value flexibility, portability, and systems of compensation that are explicitly related to their education, expertise, and job-specific performance (Clark and Monk, 2008). In many sectors, internal labor markets have been dissembled in the face of globalization (Konzelmann, 2005); embodied skills are now far less important to human resource management, even though formal educational qualifications, training, and domain-specific knowledge dominate the higher tiers of corporate employment systems (Roberts, 2004).

The modern corporation differentiates employees by performance-related compensation, while capital markets are preoccupied by short-term reported earnings (Conway et al., 2008). Consequently, there is a relationship between performance-related compensation, corporate value, and the (relatively low) value attributed by corporations to employee benefits like pensions. In this context, DB pensions are an anachronism (see Dixon and Monk, 2009 on the United States and Japan and compare with Konzelmann, 2005). There are reasonable doubts about the value of DC pensions when there is no apparent or direct link between employer provision of benefits and workers' job-specific performance deemed relevant to the reported earnings of the corporation. It is revealing that Roberts' treatise (2004) on corporate management and performance makes no mention of workplace pensions.

Where employers doubt the value of DC pensions, it is not surprising that the administrative costs of providing these benefits and the costs of matching employee contributions may be seen as the costs of doing business rather than an important aspect of employee motivation. On the other side of the employment relation, many employees see DC pension contributions as a "tax" on current income whose future value is subject to great uncertainty (a sentiment confirmed by employers who will make no commitments about what might be the "likely" long-term value of a DC pension).[1] When we asked employers "why do companies provide DC pensions?" the answers are revealing: because other companies do so; there are valuable tax advantages on contributions that accrue to higher-paid employees; and some form of supplementary pension is needed to ensure the orderly transition from work to retirement (when desired by employers).

Employer ambivalence about DC pension provision is amplified when responsibility for plan design and management is devolved to other entities, such as multi-employer plans. In many cases, these entities are often quite opaque in terms of the determination of costs and prices for even the largest participating employers. Further, the nature and quality of benefits are more often determined by reference to industry standards than the interests of participating firms. It is not surprising that DC pensions have been subject to increasing scrutiny as to the costs and the administrative burdens imposed by multi-employer plans on the companies that sponsor or participate in such plans.[2] Inevitably, outsourcing and corporate focus upon performance-related compensation have resulted in companies' human resource departments having less expertise in pension and retirement benefits than they might have had twenty years ago. As such, companies and their human resource departments may be more concerned to shift the costs of provision as well as the risks thereof to the participants, invoking arguments favoring individual autonomy and responsibility, rather than to take responsibility for apparent shortfalls in adequacy and sustainability (Jackson, 2008).

The Design of Pension Delivery

The problem of adequate workplace pensions has two sides: on one side is the lack of commitment by employers to realizing employees' retirement income aspirations; on the other side of the problem is the issue of how incentives and commitments can or should be aligned, given the apparent problems of individual decision-making noted throughout this book. This is a problem of institutional design (see Merton and Bodie, 2005). It is also, increasingly, an issue of government policy and regulation. Recognizing the ambivalence of private companies regarding the provision of pension benefits, some

---

[1] Participants' lack of confidence in the future value of DC pensions is reflected in a recent OECD (2010) report which suggested that individuals can reasonably treat future value as a "lottery" unless steps are taken to "reduce the impact of market shocks."

[2] See, for instance, the Exposure Draft of the US Financial Accounting Standards Board regarding the disclosure of employers' commitments in multi-employer retirement benefit plans (including healthcare) (Subtopic 715–80) published September 1, 2010 and available at www.fasb.org

governments have promoted alternative delivery entities, including member–profit industry consortia as well as public utilities. Courts have also become involved in this issue, because participants are typically unable to oversee their interests (if recognized as such) and the actions of the plan sponsors upon whom they rely for the realization of their retirement income objectives.[3]

In this section, we draw upon research with Roger Urwin from Towers Watson to sketch an institutional framework that could deliver workplace DC pensions in ways that would take account of both employer ambivalence and the costs of individual decision-making. This work is based upon Clark and Urwin (2011), and focuses on the logic or principles of design in the light of the apparent shortfalls in pension institutions.

The governing entity

It is axiomatic that the purpose of any governing entity of a workplace pension scheme should be the enhancement of participants' interests (the "golden" rule). Where the plan sponsor does not underwrite the final value of the pension benefit and where participants may contribute more than the plan sponsor in seeking to realize their retirement income aspirations, the governing entity should have a stake in the structure and performance of the plan (Cremers et al., 2009). There are various ways of linking the interests of the governing entity with scheme participants, including requiring the participation of entity members in the scheme and/or compensating members of the governing entity according to a priori standards of performance that include elements that match the interests of participants in cost efficiency, risk management, and long-term returns. In those institutions that take the alignment principle seriously, the governing entity is subject to fiduciary duty, combining independent members with qualified representatives of participants who have "skin in the game."

Current models of DC governance commonly rely upon single investment management firms or insurance companies selected by sponsors to provide the management and overseeing functions associated with governing boards. In these instances, those "responsible" for the pension plan typically do not have a stake in the structure and performance of the plan. In the United States,

---

[3] See the opinion of Judge Wilson in Tibble *v.* Edison International et al. CV 07-5359 (July 8, 2010) US District Court, Central District of California. But see the decision of the US District Court in Renfro *v.* Unisys Corporation et al. (No. 07-2098, April 26, 2010, p. 12) where it is asserted that "[h]aving made the decision to offer such a plan, Unisys had no incentive to waste the money that it is contributing to the plan by directing a large portion of it to a plan service provider rather than to the workers for whose benefit the plan was established. Sophisticated workers, seeing their compensation unnecessarily siphoned off to a plan administrator would ensure that workers took home a greater compensation package."

large investment organizations often take responsibility for DC delivery; in the United Kingdom, contract-based DC provision has similar issues; while in Australia it is arguable that there are moral hazard problems in both industry funds and master trusts. Where these arrangements are likely to persist, it may be necessary to use regulations to set standards, the criteria of independence, and the proper regard for participant welfare. More formally, we would suggest that the "responsible" governing entity be subject to three interrelated commitments or principles: (*a*) *decisions taken by the entity should be evaluated with respect to their implications for participants' well-being;* (*b*) *the governing entity and its members and employees together and separately are responsible for the nature, costs and performance of pension programs;* and (*c*) *the governing entity and its members and employees are accountable to plan beneficiaries.*

In stating the obvious implications of the golden rule, we have sought to resolve the conflict that sometimes appears among plan sponsors between those who decide on the nature of pension benefits offered and those who have responsibility for overseeing the implementation of benefit programs. Accountability is also to be found in transparency and disclosure, such that the governing entity has responsibility for informing beneficiaries of their policies and practices and making accessible the information necessary to evaluate the probity of board policies and practices.[4]

The executive entity

The purpose of the executive entity of a DC plan is to execute the declared principles and policies of the governing entity, whether directly through their own resources or through contracts for service written with external providers. As a matter of best practice, the executive entity are employees of the governing entity such that the CEO (and in some cases a CIO) is recruited and appointed by the governing entity and his or her actions are subject to the overseeing and approval of the governing entity. This arrangement of delegated powers is designed to resolve apparent conflicts of interest in plan sponsors where human resource departments claim responsibility for both the pension benefit "offer" as well as the purchase of pension services. It may be the case, of course, that the governing entity would wish to outsource

---

[4] Some dispute whether disclosure and transparency are effective "policing" devices, with Cain et al. (2005) arguing that disclosure may, in fact, encourage backsliding on commitments when *legitimated* by a policy of disclosure. Using experimental evidence, Church and Kuang (2009) show that disclosure combined with sanctions can be an effective policing device. Litigation, or the threat of litigation, may be a relevant sanction. Even here, though, the quality of disclosure could be more important than disclosure per se, given the costs borne by those unable to verify the quality and quantity of disclosure (see more generally Choi and Triantis, 2008).

Pension Adequacy and Sustainability

executive roles and responsibilities. In some jurisdictions, third-party providers dominate this part of the business.

This suggests, however, a well-defined "map" of decision rights and responsibilities where principles and strategy reside with the governing entity and implementation and management reside with the executive entity. Here, though, are a number of complex issues that require resolution before setting the governing structure in motion. Elsewhere, it is argued that effective decision-making depends upon time, expertise, and commitment, and that the types of decisions required can be systematically differentiated according to the timeliness and the depth of expertise and experience required, given a volatile external environment (Clark and Urwin, 2008). This argument is, of course, applicable to the relationships between the executive entity and service providers. The best institutions are explicit about delegated roles and responsibilities.

Service providers

In principle, service providers are the means by which the governing entity through its executive provides the services needed to ensure that the plan participants' goals can be realized in a cost-effective manner. Contract theory would have it that "fee for service" should be governed by time-dependent measures of performance, such that the violation of agreed measures of performance may be sanctioned in ways consistent with the significance of those violations. Here, of course, contract theory assumes that the parties on both sides of the contract are independent of one another and that rewards and sanctions are proportionate. Contract theory also assumes a competitive market for the provision of services, such that it may be cost-effective to switch between service providers given the market pricing of services or bundles of services. As such, the effective governance of service contracts depends upon the integrity of the governing entity and the independence of the executive entity.

There are, however, three connected issues which complicate the robustness of any contract model of intermediation (service provision): (*a*) Economies of scale in the financial services industry are very significant (Bikker and Dreu, 2009). DC plans are typically "small" in relation to the size of the institutions that offer services on a fee-for-service basis to sponsors and their participants. (*b*) When large service providers offer the opportunities of economies of scale, they may do so in ways that impede purchasers' capacity to oversee contract performance and their capacity to switch between service providers should agreed measures of performance be systematically flouted.[5] (*c*) Purchasers may

---

[5] Bundling can "tie" clients to their vendors by increasing the switching costs of changing vendors. Bundling can also involve cross-subsidies between services otherwise not available at a competitive price and the shrouding of prices making comparison between vendors difficult (see

find it difficult to obtain information on the true costs of individual services separate from the bundles of services on offer from service providers without the intervention of regulators.[6]

In some jurisdictions, pension plans have banded together to create their own service providers on the assumption that "shareholders" have a stronger claim on the pricing policies of service providers than "clients." Note, however, that capture is just as possible in industry consortia and service partnerships. Furthermore, there is evidence that "private" service providers may be more innovative on both the fee-for-service side of the equation as well as the quality-of-service provision than member–profit service providers owned by a number of different but cooperating pension institutions. In some cases, overlapping relationships between the members of funds' governing entities, their executive entities, and their preferred service providers may be impediments to long-term cost efficiency and the primacy of the governing entity (on behalf of participants) in relation to service providers.

Private contracts are an important but imperfect mechanism for governing the DC industry. Other institutions, including the courts, regulators, and member-based lobby groups, may also play a vital role in sustaining the accountability of governing entities and their service providers. We must be mindful of the prospect of capture and complacency as well as the need for innovation beyond current industry standards.[7]

Default, Choice, and Service

While it may be said that the *average* plan participant is neither sophisticated nor likely to appreciate the salience of saving for the future, there are likely to be those who do appreciate the significance of these issues. In fact, it would seem from our research that DC pension plans vary a great deal in terms of the

---

Gabaix and Laibson, 2006; Iacobucci, 2008). Over the long term, this can result in higher prices charged to clients as switching costs escalate in the face of entrapment. See also Shavell (2007: 325–6) on what is termed "hold-up" which "refers to situations in which a party to a new or existing contract accedes to a very disadvantageous demand, owing to the party's being in circumstances of substantial need."

[6] See the US Department of Labor's interim rule as regards the disclosure by service providers of compensation practices and possible conflicts of interest in relation to the responsibilities of ERISA plan fiduciaries; 29 CFR Part 2550, July 16, 2010.

[7] Innovation in the nature and performance of financial products is very important, given the ever-present temptation in favor of benchmarking (Clark, 2000). This practice is made possible by a lack of expertise on the buyer's side of the market and is legitimated by court decisions that use industry norms as reference points in determining whether agents have behaved properly with respect to the interests of beneficiaries. See the court opinion in Jones *v.* Harris Associates 527 F.3d 627 (7th Cir., 2008). See also the dissent led by Judge Posner where the prospect of "lower fees and higher returns" because of a commitment to best-practice in fund governance (in the mutual fund industry) is approvingly cited: Jones *v.* Harris Associates 527 F.3d 728, 731 (7th Cir., 2008).

sophistication of participants and their recognition of the salience of planning for the future. Industry affiliation appears to matter, as does the industry standing of the company sponsoring the pension plan. Furthermore, it would appear that the culture of saving varies such that country, industry, and company factors may together encourage or discourage engagement in DC schemes. In that case, a "universal" model of DC pension plan participation may not do justice to the diversity of engagement within and between sponsored pension plans. In what follows, we provide a best-practice four-step design solution to the problem of self-governance which is sensitive to the segmentation of behavior on either side of the *average* participant (as suggested in Chapter 4).

Default settings

It has become accepted practice for companies to promote participant welfare by automatically enrolling employees in the DC or DB pension plans on offer when they are first employed or when they reach some threshold in terms of hours worked and/or months employed. Not all employers are, of course, as enthusiastic about auto-enrollment as the governing entity of pension schemes (Brown, 2010). For employers, auto-enrollment normally means the take-up of their matching contributions, thereby adding to the compensation costs of the employee/participant. For a governing entity, their executive entity, and service providers, higher levels of enrollment allow a fund to reap scale economies and, in theory at least, discount the account costs borne by participants. Nonetheless, employers often use tests of eligibility for enrollment and the requirement for a deliberate, expressed choice to participate as mechanisms to screen enrollment (the cost of compensation). In some jurisdictions, legislation may be necessary to overcome employer resistance.

The other common default practice of DC plans is steering the participant to the available default fund or default strategy. While default practices vary in a number of ways, often not understood by participants, a typical strategy involves a diversified portfolio designed with reference to so-called life-cycle or target-date parameters (see Chapter 8).[8] The availability of a default strategy can help most participants who would otherwise face imperfectly understood choices among the available investment options. The advantages and limitations of this default strategy are widely acknowledged

---

[8] In a number of jurisdictions, regulators have sought to clarify what counts as a target-date fund, recognizing that there are many different versions extant in the financial services industry, and recognizing that, as a popular option, the welfare of many millions of participants is dependent upon these types of funds. See the Investor Bulletin on Target Date Retirement published by the US Department of Labor, May 6, 2010, and the intervention by the US Securities and Exchange Commission on advertising and marketing target date funds, 17 CFR Parts 230 and 270, June 23, 2010.

(compare Viceira, 2008 and Antolin et al., 2010). The most significant drawback of the default fund framework is that the only personal characteristics brought into account are age and the expected period to the target date of retirement.

In some funds, auto-enrollment and steering to the default fund are accompanied by an automatic contribution escalator that begins with a quite low contribution rate and then, over time, slowly increases to reach a contribution rate (including employers' contributions) that the governing entity deems likely to produce an "adequate" retirement income. In some funds, pension adequacy is an explicit consideration, whereas in other funds, if it is considered at all, it is done through the medium of investment policy and the like. Auto-enrollment, steering, and contribution settings are *entry-level* policy considerations and are typically justified by reference to Thaler and Sunstein's liberal paternalism (2008) (and the "safe harbor" provisions of the US Pension Protection Act, 2006; see also Pettus and Kesmodel, 2010).

More challenging is whether to allow participants to draw down from their retirement accounts for current consumption; whether to rebalance participants' asset allocations as they age and increase earned income; and whether to require annuitization at retirement or some period thereafter, such that the real value of accumulated assets is translated into a predictable and secure retirement income. Whereas it is reasonable to assume that younger employees are neither sophisticated nor appreciate the salience of saving for the future and, therefore, should be subject to *entry-level* default settings, default settings on draw-downs, rebalancing, and annuitization may be more problematic to sustain because, after twenty or thirty years of working and saving, the average participant may understand better the salience of the issue in relation to their particular circumstances.[9] In these situations, it may be that the costs for future retirement income of aberrant behavior are so significant that the governing entity should support *second-level* default settings (or the government provide legislation to protect pension adequacy).

Gates and hurdles

In fact, many DC pension plans provide mechanisms for participants to move from default settings or their initial choices to an "active choice" stance, either

---

[9] Draw-downs for current consumption are an especially troubling topic, not least because in DC plans where the participant makes the largest contribution (compared to the employer), it is arguable that account balances are properly the "property" of participants. Policies that effectively lock out the participant from access to their accumulated savings in adverse personal circumstances are an especially strong form of paternalism which some governments find difficult to justify. See Butrica et al. (2010) on the dimensions of this issue in the United States.

at will or with appropriate notification. In some best-practice funds, the environment of choice is segmented into two parts: the first part provides a limited set of options or choices selected by the governing entity (often based upon the constitutive components of the institution's default fund); and the second part provides open access to a wide range of selected products and services. By our assessment, the "guided choice" option or decision environment is less developed than the "self-selector" option or decision environment, perhaps because of the investment of resources needed to understand the particular characteristics and composition of funds' participants.

Active choice is surely the prerogative of plan participants, underwritten by liberal principles and the fact that, in many cases, accumulated assets are largely the product of individual contributions combined with investment returns. However, active choice may also be a competitive strategy of the governing entity conceived for the purpose of retaining those plan participants who carry large account balances and whose age and education are such that they recognize the salience of the issue and have confidence in their own ability (with justification or otherwise) to realize their retirement income aspirations. Retention of these participants is important because, in many cases, the overall volume of a plan's assets determines its power in the market for services and deals that may be struck to discount the quoted prices for account services, investment management, and related activities (including communication). In some cases, fewer than 5 percent of plan participants account for 50 percent or more of plan assets.

Here is a conflict of interest that goes to the heart of the governing entity's principles and policies. That is, the use of default settings, particularly those identified above as *second-level* settings, cannot be mechanisms for trapping or imprisoning plan participants whose best interests may be served in an active or controlled choice environment or, indeed, migration from the plan to other specialized service providers. In this sense, effective DC pension plans have gates that can be opened on to more sophisticated products (within the fund) and/or more sophisticated providers (outside the fund). These gates could be relatively inexpensive to open, easily accessible, and accompanied by information and decision architecture and tools (if not advice) that enhance the decision-making of the participant.

At issue, however, is whether gates should be accompanied by hurdles; that is, tests of competence and financial understanding that, in a sense, measure whether the participant is "qualified" to cope with the risks and uncertainties associated with choice in the context of accumulated savings. In this respect, the integration of gates and hurdles into the design and governance of a pension plan is, ultimately, the responsibility of the governing entity (and probably not the plan sponsor) and bears upon the unresolved issue of pension adequacy.

Engagement

The heterogeneity of many DC plans, the range of sophistication apparent among participants, and their varying levels of recognition of the importance of saving for the future suggest that best-practice pension plans have programs of engagement. At one level, engagement can be seen as an element in well-conceived strategies of retention (of participants with large account balances). At another level, engagement can be a prompt for participants to reconsider their current saving strategies in relation to future plans, thereby providing a point of intersection with the scheme and its available services. And at yet another level, engagement may satisfy those who question the legitimacy of *entry-level* and *second-level* default settings (given the presumption in favor of individual autonomy and responsibility apparent in political discourse).

Engagement policies vary by scheme, industry, and jurisdiction. Many plans offer information, websites, and briefings on current developments, retirement planning, and the range of options available to participants. It is widely noted, though, that these programs are normally sparsely attended and lack the intimacy associated with close friends and relatives dealing with the same issues. Nonetheless, if done in an informative manner, matching the consumer appeal of advertising programs aimed at specific segments of the market in evidence in the retail sector, engagement may move beyond information sharing to more effective interaction (see R. Clark et al., 2010; Goda and Flaherty, 2010). We note, however, that there are few DC sponsors who have incorporated the lessons of the behavioral revolution into the design and management of their engagement programs. As suggested in Chapter 8, "nudge" is just the tip of an iceberg which may require more radical solutions than currently contemplated (compare Thaler and Sunstein, 2008).

At the other end of the spectrum, independent financial advice provided by the fund or by the sponsor on behalf of the governing entity may attract those who appreciate the salience of the issue and have the confidence to declare their ignorance. We note, however, that this issue is the subject of debate in many countries pitting different sections of the finance industry against one another in their rush to corner what is perceived to be an expanding and lucrative market.[10] To be effective, engagement must overcome the apathy of many participants, demonstrate a level of independence rarely found in

---

[10] Ensuring participants have access to truly independent financial advisors may be an important responsibility of the governing entity of DC pension funds. This means either searching out and identifying suitable advisors and then making them available to participants at a suitable price and/or relying upon government certification programs of "independence." As such, the recently announced New Zealand government draft consultation document on a Code of Professional Conduct for Authorised Financial Advisors (March 2010) provides a template for other governments and institutions. Available at www.beehive.govt.nz

the advisory industry, and be relevant to the various circumstances of participants.

Retirement services

For many corporate sponsors, retirement has been seen as a means of jettisoning responsibility for employees' retirement saving and welfare. If consistent with the notion that DC pension plans are best treated as "arm's-length" entitlement programs, only marginally related to performance-related compensation, the governing entities of best-practice schemes have sought to retain retirees offering three types of services: information and advice on related retiree benefits; preferred providers of annuities; and techniques of budget planning and income management over the life-course. The rediscovery of retirees has been prompted by the realization that they often carry a significant portion of scheme assets, affecting the average costs borne by all participants, as well as the realization that many are vulnerable to the "independent" financial advisory industry and the costs associated with bespoke advice and services.

In some jurisdictions, the rediscovery of retirees has also been prompted by the retreat of the nation-state from providing a high level of retirement income security and ambivalence over whether retirees who rely upon so-called money-purchase schemes should purchase annuities. Many in the academic community strongly favor the purchase of annuities. However, if given the option, retirees tend not to do so (Chapter 7). In these cases, retirees' retirement income security may be better protected by their "host" pension institution than the private market for financial services. Most importantly, it is apparent that the cognitive capacity of retirees is age-related and health-related; if largely ignored in public policy, there are significant issues to be resolved about the limits of individual autonomy and responsibility in the case of onset of dementia and physical frailty (Savulescu and Hope, 2006). Whether DC plans can play a significant role in this context remains to be proven.

Reporting and communication

One aspect of the communication challenge has to do with fund reputation. It is apparent from industry respondents that DC participants respond to the clarity of reporting—the degree to which messages express clarity of purpose and the quality of execution and operation. It appears that members also respond to messages and personal leadership statements from the CEO and/ or the board. Where communication is explicitly linked to funds' reputations for quality of service, there may evolve a brand-related competitive market for pension services (as in Australia). Funds' brand strategies may be seen as part of a wider strategy of engagement by which members may be effectively immersed in the value proposition of the plan.

A second aspect of the communication challenge is related to the demonstrated efficiency and integrity of fund administration. Communication can play a significant role in giving members the confidence that their interests are paramount, and that the security of their pension savings can be assured. In effect, routinely and consistently delivering on the individual attribution of account values in the context of commingled asset pools can be a test of fund effectiveness. Considerations of tax, expenses, valuation of illiquid holdings, and timing make this a highly problematic area. Active platforms with significant daily activity have more issues to confront than those that impose less frequent dealing opportunities. Funds must balance these parameters by considering the costs and benefits of different designs for different member segments, remembering that additional functions that are valued by only a small minority segment will often entail additional costs for the majority.

Public Utilities

In sketching the elements that together represent a template for the design of DC pension institutions, we have focused upon solutions to the problems of participant competence and the ambivalence of employers to the so-called "universal" provision of pension benefits to their workforces. In doing so, we began with that which has been inherited in many jurisdictions: the trust institution with its attendant protocols as regards the proper roles and responsibilities of board members. However, we should recognize that this model is not the only model available to employers, workers, and the public at large. For example, as the rules and regulations affecting the provision of pension benefits, including DC pensions, have grown in complexity and obligations, employers have sought simpler solutions, replacing collective provision with individualized savings contracts and the like.

Another alternative to employer provision of pension benefits is to be found in public–private partnerships which combine an overarching policy framework with private institutions charged with providing the infrastructure necessary to sustain employer participation in the provision of pension benefits (Ambachtsheer, 2007). So, for example, the Australian superannuation system provides a solution to the problems of conventional DC plans through public legislation and the involvement of member–profit multi-employer industry pension plans that shift responsibility for the management and operation of the institution to an organization that mimics conventional pension plans while being subject to market competition for the enrollment of participants, their retention, and the provision of related services.[11] As such, multi-employer industry pension plans may be thought to be public utilities in the sense that their purpose is to provide benefits through the medium of regulated institutions that must negotiate their costs and prices of provision in the market for financial services.

With respect to design, the public policy framework has arguably "solved" important aspects of the problem associated with self-governance; by mandating employee and employer participation, by allowing for and regulating the default fund option, and by setting the contribution rate, the Australian government has the *entry-level* considerations that bedevil the governance of so-called voluntary systems of participation and benefit provision. Equally, by allowing for the formation and development of multi-employer institutions responsible for the purchase and provision of services consistent with high-quality DC pensions, the Federal government may have partially "solved" the problem of employer ambivalence and the attendant problems associated with conflicts of interest.

However, informed observers of the industry contend that multi-employer industry funds may not be accountable to participating companies and their employees; lack of transparency in fund decision-making combined with the high costs of switching between funds (especially significant where companies come to rely upon funds' accounting systems) may mean that participating companies are effectively locked in. At the same time, the evidence of participant-initiated switching between funds suggests that most switching is triggered by short-term differences in results between funds rather than the long-term value proposition of one fund over others.

At its core, the Australian superannuation system was built upon an assumption that competition between providers, including the financial sector, would be sufficient to ensure that the government need not regulate directly the costs charged to participants. With the regular disclosure of the costs charged by pension funds, as well as the regular disclosure of fund performance, it was hoped that the prospect of individual participants switching between funds would be sufficient to ensure a process of cost discounting, high-quality service provision, and innovation in terms of the operating frameworks and services provided to participants. The recent Cooper Review (2010: 10) of superannuation raised questions as to the effectiveness of this framework from the perspective of contributors whose welfare is particularly sensitive to the costs charged to participants. In doing so, a "plain vanilla"

---

[11] It should be acknowledged, of course, that there remain a small number of independent, corporate sponsored pension plans. As well, there are a number of large "retail" providers of pensions—what are referred to as Master Trusts. Nonetheless, the industry funds dominate the market.

"MySuper" scheme was recommended based upon low costs, a default fund rather than a choice, and a trustee-based governance regime. Here, "the (Cooper) Panel believes that by imposing some degree of homogeneity on the product, price competition might reasonably be expected to produce more positive outcomes for members and help trustees contain costs."

One implication from this recommendation is that the original framework did not take seriously the problem of participant competence: inertia reinforced by the lack of sophistication of the average participant, including the lack of awareness of the significance of minor, long-term differences in the costs of administration and the rates of performance. Another implication is that the boards of governing entities are not as independent as they might be, given apparent overlaps of board membership with industry-related consortia and membership groups. Until the publication of the Cooper Review, the government's financial regulator had financial stability as its priority, rather than apparently settled matters of fund structure and performance, including board membership, qualifications, and responsibilities.

The Cooper Review (2010: 17) also recommended that the regulator "must have a standard-setting power in relation to superannuation," thereby "overseeing and promoting industry efficiency" in a manner consistent, perhaps, with the role and responsibilities of the UK Pensions Regulator. Though path-breaking when conceived, the Australian system has yet to realize its potential and provide solutions to the scope of issues we have identified as consistent with best-practice DC pension fund governance.

By contrast, the UK government's NEST (National Employment Savings Trust) scheme seeks to take advantage of the lessons learned from the Australian experience to provide a public utility of a different kind. In this case, instead of making participation mandatory, the government has required employers to implement auto-enrollment systems leaving open the option for employers to provide their own superior pension systems as well as the option for individuals to opt out entirely from this form of pension saving. Like the Australian system, the UK government has set a combined minimum contribution rate: in this case 8 percent of gross salary, falling rather short of the Australian contribution rate.[12] In the UK case, moreover, the UK government will sponsor a "universal" default fund that together will collect perhaps as many as six million plan participants, leaving those inclined to make their own arrangements to either their employer's plan or the market for private pensions. The government plan is designed to reap economies of scale on the

---

[12] In the May 11, 2010 federal budget statement, the Australian government indicated that mandatory contributions will rise to 12 percent of gross earnings, with the government providing a contribution on behalf of low-income earners.

costs of pension provision, including the purchase of services from the global financial industry.

The Australian system is a partial solution to the problem of pension adequacy, in that it has not yet engaged with the issues we identify as *second-level* elements of design. The proposed UK system has yet to indicate its commitment to the *second-level* agenda. Furthermore, it remains to be seen whether a universal public utility would function better than a relatively decentralized system of provision subject to market competition. There are also significant governance issues involved when a single government-sponsored pension plan carries the retirement income aspirations of so many relatively lower-paid individuals. While recognizing the ambitious low-cost framework that NEST will use as a competitive strategy, reconciling its ambitions to provide a secure and affordable pension with the possible costs of market-following by simple passive index funds (the OECD's 2010 "lottery") and the rather low rates of mandated contributions remains a formidable challenge. These issues may conspire to limit its effectiveness when judged against the retirement needs of low-income participants.

A People's Pension

If the previous section focused on making existing pension institutions, in particular DC pension plans, better—fairer, more responsive, more cost-effective—it remains true that the public utility model only partly addresses the fundamental agency problem. If firms do not perceive employee pensions to be in their best interests (in terms of staff recruitment, retention, and management) and if pensions are seen as a cost and a burden in terms of the firm's valuation, firms will continue to minimize that burden by reducing costs (notwithstanding inertia and tradition; see Bridgen and Meyer, 2005) and ensuring that the maximum amount of risk will be borne by the individual and the state. In such cases, the rhetoric of choice and responsibility is mere cover for the individualization of risk.

What this book, and other research on financial decision-making and literacy, has illustrated is that cognitive skills related to the social capital of economic decision-making are not evenly distributed (Atkinson et al., 2007; Bernheim, 1998; Cutler, 1997; Lusardi and Mitchell, 2005, 2008). These skills are correlated with class, as defined by income and occupation, as well as age, gender, and marital status; there is also a possible spatial association in terms of different geographically embedded communities of practice and ecologies of finance (Leyshon et al., 2004). More work is needed to conceptualize and operationalize the notion of context in ways that can account for the complex interplay of these factors, the role of interpersonal relationships, and the impacts of structural constraints (see Strauss, 2009b for a suggested approach). Models of behavior that acknowledge the limits to rationality, tend not to go beyond documenting and theorizing limits to human reasoning and focus (rightly) on the issue of plan design, leaving aside the risks allocated to those least able, or willing, to bear it.

We would like to conclude by returning to the issues of adequacy and sustainability in the context of the public utility model, and to make some points about what a pension model might look like—one that takes the lessons of behavioralism seriously. We call this "a people's pension," for it is based on the broad conception of personhood that underpins our book. If we look at the issue of adequacy, it is clear that a pension model that shifts the risks of pension value onto all individuals turns plan members into de facto trustees. It accepts that employers have little to contribute in terms of pension value or commitment with the implication that, in combination, such a model will likely not significantly increase the welfare of those on low incomes.

This is an issue of concern in many jurisdictions: those women in particular who work part-time and take career breaks, and those from ethnic minorities who are the most likely to be in precarious and temporary jobs, are unlikely to build up an adequate pension pot and are likely the least able to exercise control in terms of active choices. While NEST embodies some of the best characteristics of a public utility model, and the benefits of more flexible, portable pensions, it also highlights the drawbacks of the DC model itself: relatively high initial charges, a lack of commitment from employers, and a lack of security in terms of the income it will provide. There is every chance that NEST and other similar institutions will fail the adequacy test for the low paid; it is unlikely to enhance their incomes significantly especially considering the role that the price of annuities inevitably plays in the equation, and the uncertainty associated with means-testing. This effect could be particularly pernicious if NEST serves to cover government discounting of the BSP (in the aftermath of the global financial crisis).

We are also doubtful about the sustainability of such a solution. It may be affordable, but, if it fails to increase the incomes of the low paid in retirement, it will not be socially sustainable. Moreover, there is a chance that the introduction of NEST will accelerate the pension gap between those who have more than minimal coverage and those who rely solely on NEST. There is evidence that DB pensions in the private sector have become the preserve of the powerful (senior executives), the well-qualified, the well-paid, and those in legacy industries; the public sector is likely to follow suit in the long run. It is particularly telling that the same companies that are closing DB schemes to new members and new accruals often retain "gold-plated" pensions for senior executives (Inman, 2009). If this trend continues to accelerate, it will compound the increasing polarization in incomes and wealth. This, as much as

Pension Adequacy and Sustainability

the breakdown in intergenerational solidarity that some claim high pension costs will produce, is a threat to social cohesion.

In Chapter 1, we noted the fundamental social, economic, and cultural shifts that undermined the tripartite institutions of the Keynesian welfare state, the standard contract of employment, and the male breadwinner-headed nuclear family. To make good on the virtues of behavioralism, these broader issues need to be taken into account: these issues are representative of the social processes that structure "the field" (to use Bourdieu's 2005 term) in which financial decision-making takes place. Thus, a people's pension must also recognize that the institutional landscape has changed. We would argue that this presents a stronger, rather than a weaker, argument for the collectivization of risk in the domain of retirement saving. So, for example, a people's pension for the low income would rely upon an enhanced state pension, underwriting the benefits due to pension savings (whether in a NEST-like institution or through other employer options), where tax policy enhances people's long-term savings behavior rather than penalizing participation, and where "choice" reinforces commitment rather than carrying the risks of a "lottery." Inevitably, pension policy may have to contain a significant mandatory component.

This is not to deny the right of people to exercise choice. But choice in retirement planning, we would argue, can be exercised in supplementary pension schemes that are designed in ways to manage the downside of choice-gone-wrong. Those with the assets and appetite for risk can invest as they wish, subject to appropriate governance, regulatory, and tax regimes. To burden all workers with the risk and responsibility for assuring an adequate income in retirement, without guaranteeing adequacy, is to place enormous faith in their decision-making abilities and propensities, their social capital, and their innate economic rationality.

Bibliography

Abolafia, M. (1996), *Making Markets: Opportunism and Restraint on Wall Street*, Cambridge, MA: Harvard University Press.

Adlrich, J. H. and Nelson, F. D. (1983), 'Linear probability, logit, and probit models', Sage University Paper Series on Quantitative Applications in the Social Sciences, 07–045, London: Sage.

Agarwal, S., Driscoll, J., Gabaix, X., and Laibson, D. (2008), 'The age of reason: financial decisions over the lifecycle', available online at: http://papers.ssrn.com/sol3/papers.cfm?abstract_id=973790.

Ainslie, G. (2001), *Breakdown of Will*, Cambridge: Cambridge University Press.

—— (2005), Précis of *Breakdown of Will*, *Behavioral and Brain Sciences*, 28: 635–50.

Akerlof, G. A. and Shiller, R. J. (2009), *Animal Spirits: How Psychology Drives the Economy, and Why it Matters for Global Capitalism*, Princeton: Princeton University Press.

Allen, F. and Gale, D. (2007), *Understanding Financial Crises*, Oxford: Oxford University Press.

Ambachtsheer, K. (2007), *Pension Revolution: A Solution to the Pensions Crisis*, New York: John Wiley.

Amemiya, T. (1981), 'Qualitative response models: A survey', *Journal of Economic Literature*, 19: 1483–536.

Ameriks, J. and Mitchell, O. S. (eds.) (2008), *Recalibrating Retirement Spending and Saving*, Oxford: Oxford University Press.

Antolin, P., Payet, S., and Yermo, J. (2010), 'Assessing Default Investment Strategies in Defined Contribution Pension Plans', Working Paper No. 2, Paris: OECD.

Armstrong, M., Cowan, S., and Vickers, J. (1998), *Regulatory Reform: Economic Analysis and the British Experience*, Cambridge, MA: MIT Press.

Arrow, K. J. (1986), 'Rationality of self and others in an economic system', *Journal of Business*, 59: S385–99.

Astuti, R. and Harris, P. L. (2008), 'Understanding mortality and the life of the ancestors in rural Madagascar', *Cognitive Science*, 32: 713–40.

Atiyah, P. S. (1979), *The Rise and Fall of Freedom of Contract*, Oxford: Clarendon Press.

Atkinson, A. (2008), *Evidence of impact: an overview of financial education evaluations*, Consumer Research Report 68, London: Financial Services Authority.

——, McKay, S., Collard, S., and Kempson, E. (2007), 'Levels of Financial Capability in the UK', *Public Money and Management*, 27(1): 29–36.

Bibliography

Audi, R. (2007), *Moral Value and Human Diversity*, Oxford: Oxford University Press.

Axelrod, R. (1984), *The Evolution of Cooperation*, New York: Basic Books.

Bajtelsmit, V. L. (2006), 'Gender, the family, and economy', in Clark, G. L., Munnell, A. H., and Orszag, J. M. (eds.), *The Oxford Handbook of Pensions and Retirement Income* (pp. 121–40), Oxford: Oxford University Press.

—— and Bernasek, A. (1996), 'Why do women invest differently than men?', *Financial Counselling and Planning*, 7: 1–10.

—— —— and Jianakoplos, N. A. (1999), 'Gender differences in defined contribution pension decisions', *Financial Services Review*, 8: 1–10.

—— and Jianakoplos, N. (2000), 'Women and pensions: a decade of progress?', EBRI Issue Brief, Washington, DC.

Banks, J., Blundell, R., Oldfield, Z., and Smith, J. P. (2007), 'Housing Price Volatility and Downsizing in Later Life', NBER Working Paper No. 13496, Cambridge, MA: National Bureau of Economic Research.

Bao, Y. Q., Zhou, K. Z., and Su, C. (2003), 'False consciousness and risk aversion: do they affect consumer decision-making?', *Psychology and Marketing*, 20: 733–55.

Barnes, T. J. (1988), 'Rationality and Relativism in Economic-Geography - an Interpretive Review of the Homo Economicus Assumption', *Progress in Human Geography*, 12(4): 473–96.

Baron, J. (2008), *Thinking and Deciding*, 4th edn, Cambridge: Cambridge University Press.

Barro, R. (2006), 'Rare disasters and asset markets in the twentieth century', *Quarterly Journal of Economics*, 114: 823–64.

Bathelt, H. (2006), 'Geographies of production: growth regimes in spatial perspective 3 - toward a relational view of economic action and policy', *Progress in Human Geography*, 30: 223–36.

—— and Glückler, J. (2005), 'Resources in economic geography: from substantive concepts to a relational perspective', *Environment and Planning A*, 37: 1545–63.

Baumeister, R. F. (2004), 'The psychology of irrationality: why people make foolish, self-defeating choices', in Brocas, I. and Carrillo, J. D. (eds.), *The Psychology of Economic Decisions. Volume 1: Rationality and Well-Being* (pp. 3–16), Cambridge: Cambridge University Press.

BBC (2003), 'Dow Jones milestones', retrieved 25 June 2006, from: http://news.bbc.co.uk/1/hi/business/business_basics/145986.stm.

Beck, U. and Beck-Gernsheim, E. (2001), *Individualization*, London: Sage.

Bellamy, K. and Rake, K. (2005), *Money, Money, Money: Is it Still a Rich Man's World?*, London: The Fawcett Society.

Benartzi, S. and Thaler, R. (2001), 'Naive diversification strategies in defined contribution savings plans', *American Economic Review*, 91: 71–99.

—— —— (2002), 'How much is investor autonomy worth?', *Journal of Finance*, 57: 1593–616.

—— —— (2005), 'Save more tomorrow: using behavioral economics to increase employee savings', *Journal of Political Economy*, 112: 164–87.

Bermúdez, J. L. (2009), *Decision Theory and Rationality*, Oxford: Oxford University Press.

Bibliography

Bernasek, A. and Shwiff, S. (2001), 'Gender, risk and retirement', *Journal of Economic Issues*, 35: 345–56.

Bernheim, D. (1998), 'Financial illiteracy, education and retirement saving', in Mitchell, O. S. and Schieber S. (eds.), *Living with Defined Contribution Pensions* (pp. 38–68), Philadelphia, PA: University of Pennsylvania Press.

Bertrand, M., Karlan, D., Mullainathan, S., Shafir, E., and Zinman, J. (2005), 'What's psychology worth? A field experiment in the consumer credit market', Working Paper No. 11892, Cambridge, MA: National Bureau of Economic Research.

Bikker, J. and Dreu, J. (2009), 'Operating Costs of Pension Funds: The impact of scale, governance and plan design', *Journal of Pension Economics and Finance*, 8: 63–89.

Bodie, Z. and Treussard, J. (2007), 'Perspectives: making investment choices as simple as possible, but not simpler', *Financial Analysts Journal*, 63(6): 42–7.

Borio, C. (2006), 'Monetary and financial stability: here to stay?', *Journal of Banking and Finance*, 30: 3407–14.

Bourdieu, P. (2005), *The Social Structures of the Economy*, Cambridge: Polity Press.

Boyer, R. (2000), 'Is a finance-led growth regime a viable alternative to Fordism? A preliminary analysis', *Economy and Society*, 29: 111–45.

Bratman, M. E. (1987), *Intention, Plans, and Practical Reason*, Cambridge, MA: Harvard University Press.

—— (2007), *Structures of Agency: Essays*, Oxford: Oxford University Press.

—— (2009), 'Intention, practical rationality, and self-governance', *Ethics*, 119: 411–43.

Bridgen, P. and Meyer, T. (2005), 'When Do Benevolent Capitalists Change Their Mind? Explaining the Retrenchment of Defined-benefit Pensions in Britain', *Social Policy and Administration*, 39(7): 764–85.

Brown, S. K. (2010), *Automatic 401(k) Plans: Employer Views on Enrolling New and Existing Employees*, Washington, DC: AARP Research and Strategic Analysis.

Butrica, B. A., Zedlewski, S.R., and Issa, P. (2010), 'Understanding Early Withdrawals From Retirement Accounts', Discussion Paper 10–02, Washington, DC: Urban Institute.

Byrne, A., Blake, D. P., Dowd, K., and Cairns, A. J. G. (2007), 'Default funds in UK DC plans', *Financial Analysts Journal*, 63(4): 40–51.

Cain, D. M., Loewenstein, G., and Moore, D. (2005), 'The dirt on coming clean: perverse effects of conflicts of interest', *Journal of Legal Studies*, 34: 1–25.

Callon, M. (1998), *The Laws of the Markets*, Oxford: Blackwell.

Camerer, C. (2008), 'The potential of neuroeconomics', *Economics and Philosophy*, 24: 369–79.

Cannon, E. and Tonks, I. (2004), 'UK annuity price series 1957–2002', *Financial History Review*, 2: 165–96.

—— —— (2005), *Survey of annuity pricing*, Research Report 318, Department of Work and Pensions, London: The Stationery Office.

Carroll, C. D. (2000), 'Portfolios of the rich', Working Paper No. 7826, Cambridge, MA: National Bureau of Economic Research.

Caudill, S. (1988), 'An advantage of the linear probability model over probit or logit', *Oxford Bulletin of Economics and Statistics*, 50: 425–7.

Charles, N. and Harris, C. (2007), 'Continuity and change in work-life balance choices', *British Journal of Sociology*, 58(2): 277–95.

Choi, J. J., Laibson, D., and Madrian, B. C. (2004), 'Plan design and 401(k) savings outcomes', *National Tax Journal*, 57: 275–98.

—————— (2005), '$100 bills on the sidewalk: suboptimal saving in 401(k) plans', Working Paper No. 11554, Cambridge, MA: National Bureau of Economic Research.

—————— and Metrick, A. (2002), 'Defined contribution pensions: plan rules, participant decisions, and the path of least resistance', in J. M. Poterba (ed.), *Tax Policy and the Economy*, Vol. 16 (pp. 67–113), Cambridge, MA: MIT Press.

Choi, A. and Triantis, G. (2008), 'Completing contracts in the shadow of costly verification', *Journal of Legal Studies*, 37: 503–34.

Christie, H., Smith, S. J., and Munro, M. (2008), 'The emotional economy of housing', *Environment and Planning A*, 40: 2296–312.

Church, B. K. and Kuang, X. (2009), 'Conflicts of interest, disclosure, and (costly) sanctions: experimental evidence', *Journal of Legal Studies*, 38: 505–32.

Clark, G. L. (2000), *Pension Fund Capitalism*, Oxford: Oxford University Press.

—— (2003), *European Pensions and Global Finance*, Oxford: Oxford University Press.

—— (2006a), 'The UK occupational pension system in crisis', in Pemberton, H., Thane, P., and Whiteside, N. (eds.), *Britain's Pensions Crisis: History and Policy* (pp. 145–68), London: Oxford University Press for the British Academy.

—— (2006b), 'The regulation of pension fund governance', in Clark, G. L., Munnell, A., and Orszag, M. (eds.), *The Oxford Handbook of Pensions and Retirement Income* (pp. 483–500), Oxford: Oxford University Press.

—— (2008), 'Governing finance: global imperatives and the challenge of reconciling community representation with expertise', *Economic Geography*, 84: 281–302.

—— (2010), 'Human nature, the environment, and behaviour: explaining the scope and geographical scale of financial decision-making', *Geografiska Annaler: B, Human Geography*, 92(2): 159–73.

——, Caerlewy-Smith, E., and Marshall, J. C. (2006), 'Pension fund trustee competence: decision making in problems relevant to investment practice', *Journal of Pension Economics and Finance*, 5(1): 91–110.

—————— (2007), 'The consistency of UK pension fund trustee decision-making', *Journal of Pension Economics and Finance*, 6(1): 67–86.

—————— (2009), 'Solutions to the asset allocation problem by informed respondents: the significance of the size-of-bet and the 1/n heuristic', *Risk Management and Insurance Review*, 12(2): 251–71.

——, Duran-Fernandez, R., and Strauss, K. (2010), '"Being in the market": the UK house-price bubble and the intended structure of individual pension investment portfolios', *Journal of Economic Geography*, 10(3): 331–59.

——, Feldman, M., and Gertler, M. S. (eds.) (2000), *The Oxford Handbook of Economic Geography*, Oxford: Oxford University Press.

—— and Knox-Hayes, J. (2007), 'Mapping UK pension benefits and the intended purchase of annuities in the aftermath of the 1990s stock market bubble', *Transactions of the Institute of British Geographers*, NS32(4): 539–55.

Bibliography

Clark, G. L., and Knox-Hayes, J. (2009), 'The "new" paternalism, consultation and consent: expectations of UK participants in defined contribution and self-directed retirement savings schemes', *Pensions: An International Journal*, 14(1): 58–74.

—— —— and Strauss, K. (2009), 'Financial sophistication, salience, and the scale of deliberation in UK retirement planning', *Environment and Planning A*, 41(10): 2496–515.

—— and Monk, A. H. B. (2007), 'The "crisis" in defined benefit corporate pension liabilities', *Pensions: An International Journal*, 12(1): 43–54 and 12(2): 68–81.

—— —— (2008), 'Conceptualizing the defined benefit pension promise', *Benefits Quarterly*, 24(1): 7–18.

——, Munnell, A. H., and Orszag, J. M. (eds.) (2006), *The Oxford Handbook of Pensions and Retirement Income*, Oxford: Oxford University Press.

—— and O'Connor, K. (1997), 'The informational content of financial products and the spatial structure of the global financial industry', in Cox, K. R. (ed.), *Spaces of Globalization: Reasserting the Power of the Local* (pp. 89–144), New York: Guilford Press.

—— and Strauss, K. (2008), 'Individual pension-related risk propensities: the effects of socio-demographic characteristics and a spousal pension entitlement on risk attitudes', *Ageing and Society*, 28: 847–74.

——, Thrift, N., and Tickell, A. (2004), 'Performing finance: the industry, the media and its image', *Review of International Political Economy*, 11: 289–310.

—— and Urwin (2008), 'Making pension boards work: the critical role of leadership', *Rotman Journal of International Pension Management*, 1: 38–45.

—— —— (forthcoming 2011), 'DC Pension Fund Best-practice Design and Governance', *Benefits Quarterly*.

—— and Whiteside, N. (2003), 'Introduction' in Clark, G. L. and Whiteside, N. (eds.), *Pension Security in the 21st Century: Redrawing the Public-Private Debate* (pp. 1–20), Oxford: Oxford University Press.

—— and Wójcik, D. (2007), *The Geography of Finance: Corporate Governance in the Global Marketplace*, Oxford: Oxford University Press.

—— and Wrigley, N. (1997), 'Exit, the firm and sunk costs: re-conceptualising the corporate geography of disinvestment and plant closure', *Progress in Human Geography*, 21: 338–58.

Clark, R. L., Morrill, M. S., and Allen, S. G. (2010), 'Employer-provided Retirement Planning Programs', in Clark, R. L. and Mitchell, O. S. (eds.), *Reorienting Retirement Risk Planning* (pp. 36–63), Oxford: Oxford University Press.

Compton, J. and Pollack, R. A. (2007), 'Why are power couples increasingly concentrated in large metropolitan areas?', *Journal of Labor Economics*, 25: 475–512.

Conefrey, T. and Gerald, J. F. (2009), *Blowing Bubbles – and Bursting Them: The Case of Ireland and Spain*, Paris: OECD.

Conway, N., Deakin, S., Konzelmann, S., Petit, H., Rebérioux, A., and Wilkinson, F. (2008), 'The Influence of Stock Market Listing on Human Resource Management: Evidence for France and Britain', *British Journal of Industrial Relations*, 46: 631–73.

Cooper Review (2010), *Super System Review: Final Report*, Canberra: Commonwealth of Australia.

Costa, D. L. and Kahn, M. E. (2000), 'Power couples: changes in the locational choice of the college educated, 1940–1990', *The Quarterly Journal of Economics*, 115(4): 1287–315.

Couclelis, H. (2009), 'Rethinking time geography in the information age', *Environment and Planning A*, 41(7): 1556–75.

Coval, J. D. and Moskowitz, T. J. (2001), 'The geography of investment: informed trading and asset prices', *Journal of Political Economy*, 109: 811–41.

Cremers, M., Driessen, J., Maenhout, P., and Weinbaum, D. (2009), 'Does Skin in the Game Matter? Director Incentives and Governance in the Mutual Fund Industry', *Journal of Financial and Quantitative Analysis*, 44: 1345–73.

Cronqvist, H. and Thaler, R. (2004), 'Design choices in privatized social-security systems: learning from the Swedish experience', *American Economic Review*, 94: 424–8.

Cutler, N. E. (1997), 'The false alarms and blaring sirens of financial literacy: middle-agers' knowledge of retirement income, health finance, and long-term care', *Generations-Journal of the American Society on Aging*, 21(2): 34–40.

Cyert, R. and March, J. (1963), *A Behavioral Theory of the Firm*, Oxford: Blackwell.

Damagio, A. (2003), *Looking for Spinoza*, London: Vintage.

Damasio, A. R. (1994), *Descartes' Error: Emotion, Reason and the Human Brain*, New York: Putnam.

Damasio, H., Grabowski, T., Frank, R., Galaburda, A., and Damasio, A. R. (1994), 'The return of Phineas Gage: clues about the brain from the skull of a famous patient', *Science*, 264: 1102–5.

Deaton, A. (1992), *Understanding Consumption*, Oxford: Oxford University Press.

De Deken, J. J., Ponds, E., and van Riel, B. (2006), 'Social solidarity', in Clark, G. L., Munnell, A. H., and Orszag, J. M. (eds.), *The Oxford Handbook of Pensions and Retirement Income* (pp. 141–60), Oxford: Oxford University Press.

Department of Social Security (DSS) (1998), *Partnership in 'Pensions'*, Cm 4179, London: The Stationery Office.

Department for Work and Pension (DWP) (1998), *A new contract for welfare: partnership in pensions*, London: Department for Work and Pensions.

—— (2006), *Personal Accounts: A New Way to Save*, Cm 6975, London: The Stationery Office.

Dickens, R. (2000), 'The evolution of individual male earnings in Great Britain: 1975–95', *Economic Journal*, 110: 27–49.

Disney, R., Henley, A., and Stears, G. (2002), 'Housing costs, house price shocks and savings behaviour among older households in Britain', *Regional Science and Urban Economics*, 32(5): 607–25.

Dixon, A. D. and Monk, A. H. B. (2009), 'The power of finance: accounting harmonization's effect on pension provision', *Journal of Economic Geography*, 9(5): 619–39.

Doherty, M. E. (2003), 'Optimists, pessimists, and realists', in Schneider, S. L. and Shanteau, J. (eds.), *Emerging Perspectives on Judgement and Decision Research* (pp. 643–79), Cambridge: Cambridge University Press.

Bibliography

Dohmen, T., Falk, A., Huffman, D., Sunde, U., Schupp, J., and Wagner, G. G. (2005), 'Individual risk attitudes: new evidence from a large, representative, experimentally-validated survey', Discussion Paper 1730, Bonn: Institute for the Study of Labor, available online at: ftp://repec.iza.org/RePEc/Discussionpaper/dp1730.pdf.

Dolvin, S. D. and Templeton, W. K. (2006), 'Financial education and asset allocation', *Financial Services Review*, 15: 133–49.

Dorn, D. and Huberman, G. (2005), 'Talk and action: what individual investors say and what they do', *Review of Finance*, 9: 437–81.

Døskeland, T. M. and Hvide, H. K. (2011), 'Do individual investors have asymmetric information based on work experience?', *Journal of Finance*, 66(3): 1011–41.

Douglas, M. and Wildavsky, A. (1983), *Risk and Culture: An Essay on the Selection of Technological and Environmental Dangers*, Berkeley: University of California Press.

Dworkin, G. (1972), 'Paternalism', *The Monist*, 56: 64–84.

Easterlow, D. and Smith, S. J. (2004), 'Housing for health: can the market care?', *Environment and Planning A*, 36: 999–1017.

Ebbinghaus, B. (2006), *Reforming Early Retirement in Europe, Japan and the USA*, Oxford: Oxford University Press.

—— (2010), *Varieties of Governing Pensions*, Oxford: Oxford University Press.

Engelen, E. (2003), 'The logic of funding European pension restructuring and the dangers of financialization', *Environment and Planning A*, 35: 1357–72.

Esping-Andersen, G. (1989), 'The Three Political Economies of the Welfare State', *Canadian Review of Sociology and Anthropology*, 26(1): 10–36.

—— (1999), *Social Foundations of Postindustrial Economies*, Oxford: Oxford University Press.

Foster, L. (2010), 'Towards a new political economy of pensions? The implications for women', *Critical Social Policy*, 30(1): 27–47.

Foucault, M. (1977), *Discipline and Punish: The Birth of the Prison*, A. Sheridan (trans.), London: Allen Lane.

—— (1979), *The History of Sexuality*, R. Hurley (trans.), London: Allen Lane.

Frericks, P., Maier, R., and De Graaf, W. (2007), 'European pension reforms: Individualization, privatization and gender pension gaps', *Social Politics*, 14(2): 212–37.

Friedberg, L. and Webb, A. (2006), 'Determinants and consequences of bargaining power in households', Working Paper No. 12367, Cambridge, MA: National Bureau of Economic Research.

Fries, J. F. (1980), 'Aging, natural death and the compression of morbidity', *New England Journal of Medicine*, 303: 130–35.

Gabaix, X. and Laibson, D. (2006), 'Shrouded attributes, consumer myopia, and information suppression in competitive markets', *Quarterly Journal of Economics*, 113: 505–40.

—— ——, Moloche, G., and Weinberg, S. (2006), 'Costly information acquisition: experimental analysis of a boundedly rational model', *American Economic Review*, 96: 1043–68.

Gardner, J. and Orszag, J. M. (2004), 'How have older workers responded to scary markets?', Technical Paper 2003-TR-07, London: Watson Wyatt, available online at: http://ssrn.com/abstract=892764.

Gertler, M. S. (2001), 'Best practice? Geography, learning and the institutional limits to strong convergence', *Journal of Economic Geography*, 1: 5–26.

—— (2003), 'Tacit knowledge and the economic geography of context, or the undefinable tacitness of being (there)', *Journal of Economic Geography*, 3: 75–99.

Ghilarducci, T. (2008), *When I'm Sixty-Four: The Plot Against Pensions and the Plan to Save Them*, Princeton: Princeton University Press.

Gigerenzer, G. (2004), 'The irrationality paradox', *Brain and Behavioral Sciences*, 27: 336–8.

——, Todd, P. M., and the ABC Research Group (1999), *Simple Heuristics That Make Us Smart*, New York: Oxford University Press.

Ginn, J. (2003), *Gender, Pensions and the Lifecourse: How Pensions Need to Adapt to Changing Family Forms*, Bristol: Policy Press.

—— and Arber, S. (1991), 'Gender, class and income inequalities in later life', *British Journal of Sociology*, 42: 369–96.

——, Street, D., and Arber, S. (2001), *Women, work and pensions: international issues and prospects*, Buckingham: Open University Press.

Glennie, P. and Thrift, N. (2005), 'Revolutions in the times. Clocks and the temporal structures of everyday life', in Livingstone, D. and Withers, C. (eds.), *Geography and Revolution* (pp. 160–98), Chicago: University of Chicago Press.

—— —— (2009), *Shaping the Day: A History of Timekeeping in England and Wales 1300–1800*, Oxford: Oxford University Press.

Goda, G. S. and Flaherty, C. (2010), 'Incorporating Employee Heterogeneity into Default Rules for Retirement Plan Selection', Chestnut Hill, MA: Boston College Center for Retirement Research Working Paper No. 2010–6, available online at: http://ssrn.com/abstract=1611782.

Goffman, E. (1974), *Stigma: Notes on the Management of Spoiled Identity*, New York: J. Aronson.

Goldstein, W. M., Barlas, S., and Beattie, J. (2001), 'Talk of tradeoffs: judgements of relative importance', in Weber, E. U., Baron, J., and Loomes, G. (eds.), *Conflict and Tradeoffs in Decision Making* (pp. 175–203), Cambridge: Cambridge University Press.

Goldsticker, R. (2007), 'A mutual fund to yield annuity-like benefits', *Financial Analysts Journal*, 63(1): 63–7.

Goos, M. and Manning, A. (2007), 'Lousy and lovely jobs: the rising polarization of work in Britain', *Review of Economics and Statistics*, 89: 118–33.

Granovetter, M. (1985), 'Economic action and social structure: the problem of embeddedness', *American Journal of Sociology*, 91: 481–510.

Gross, D. M. (2006), *The Secret History of Emotion: From Aristotle's Rhetoric to Modern Brain Science*, Chicago: University of Chicago Press.

Hägerstrand, T. (1970), 'What about people in human geography?', *Papers, Regional Science Association*, 24: 7–21.

Hallahan, T. A., Faff, R. W., and McKenzie, M. D. (2004), 'An empirical investigation of personal financial risk tolerance', *Financial Services Review*, 13: 57–78.

Hamnett, C. (1999), *Winners and Losers: Home Ownership in Modern Britain*, London: Routledge.

Bibliography

Hamnett, C. (2009), 'Spatially displaced demand and the changing geography of house prices in London, 1995–2006', *Housing Studies*, 24: 301–20.

—— and Whitelegg, C. (2007), 'Loft conversion and gentrification in London: from industrial to post-industrial land use', *Environment and Planning A*, 39: 106–24.

Harré, R. (1993), *Social Being*, 2nd edn, Oxford: Blackwell.

Heaton, J. and Lucas, D. (2000), 'Portfolio choice in the presence of background risk', *Economic Journal*, 110: 1–26.

Henrich, J., Boyd, R., Bowles, S., Camerer, C., Fehr, E., Gintis, H., McElreath, R., Alvard, M., Barr, A., Ensminger, J., Smith Henrich, N., Hill, K., Gil-White, F., Gurven, M., Marlowe, F. W., Patton, J. Q., Tracer, D. (2005), '"Economic man" in a cross-cultural perspective: behavioural experiments in 15 small-scale societies', *Behavioral and Brain Sciences*, 28: 795–815.

Hershey, D., Henkens, K., and van Dalen, H. P. (2008), 'Aging and retirement planning: interdisciplinary influences viewed through a cross-cultural lens', Working Paper, Stillwater, OK: Department of Psychology, Oklahoma State University.

Hills, J. (2010), *An Anatomy of Economic Inequality in the UK*, London: HM Government Equalities Office.

Hilton, D. J. (2003), 'Psychology and the financial markets: applications to understanding and remedying irrational decision-making', in Brocas, I. and Carrillo, J. D. (eds.), *The Psychology of Economic Decisions. Volume 1: Rationality and Well-Being* (pp. 273–97), Oxford: Oxford University Press.

HM Government (2007), *Homes for the Future: More Affordable, More Sustainable*, Department for Communities and Local Government Cm 7191, London: The Stationery Office.

Hogarth, R. M. (2001), *Educating Intuition*, Chicago: University of Chicago Press.

Hollis, M. (1996), *Reason in Action: Essays in the Philosophy of Social Science*, Cambridge: Cambridge University Press.

Holmes, S. (1993), *The Autonomy of Antiliberalism*, Cambridge, MA: Harvard University Press.

Hong, H., Kubik, J., and Stein, J. (2004), 'Social interaction and stock-market participation', *Journal of Finance*, 59: 137–63.

—— —— —— (2005), 'Thy neighbor's portfolio: word-of-mouth effects in the holdings and trades of money managers', *Journal of Finance*, 60: 2801–24.

Huberman, G. (2000), 'Home bias in equity markets: international and intranational evidence', in Hess, G. D. and van Wincoop, E. (eds.), *Intranational Macroeconomics* (pp. 76–91), Cambridge: Cambridge University Press.

—— (2001), 'Familiarity breeds investment', *Review of Financial Studies*, 14: 659–80.

—— (2003), 'Behavioral finance and markets', in Dimitri, N., Basili, M., and Gilboa, I. (eds.), *Cognitive Processes and Economic Behaviour* (pp. 1–14), London: Routledge.

Hurley, S. (2008), 'The shared circuits model (SCM): how control, mirroring, and simulation can enable imitation, deliberation, and mindreading', *Behavioral and Brain Sciences*, 31: 1–21.

Iacobucci, E. (2008), 'A switching costs explanation of tying and warranties', *Journal of Legal Studies*, 37: 431–58.

Inman, P. (2009), 'Executive pensions: how Sir Fred Goodwin's gold-plated package fits in', *The Guardian*, London, Friday 27 February 2009.

Iyengar, S. (2010), *The Art of Choosing*, New York: Hachette Book Group.

—, Jiang, W., and Huberman, G. (2004), 'How much choice is too much: determinants of individual contributions in 401K retirement plans', in Mitchell, O.S. and Utkus, S. P. (eds.), *Pension Design and Structure: New Lessons from Behavioral Finance* (pp. 83–97), Oxford: Oxford University Press.

Jackson, H. (2008), 'The trilateral dilemma in financial regulation', in Lusardi, A. (ed.), *Overcoming the Saving Slump* (pp. 82–117), Chicago: University of Chicago Press.

Jacobs-Lawson, J. M. and Hershey, D. A. (2005), 'Influence of future time perspective, financial knowledge, and financial risk tolerance on retirement saving behaviours', *Financial Services Review*, 14: 331–44.

JRF (2010), *Can equity release help older home-owners improve their quality of life?*, York: Joseph Rowntree Foundation.

Kahneman, D. (2003), 'Maps of bounded rationality: psychology for behavioral economics', *American Economic Review*, 93(5): 1449–75.

—, Schwartz, A., Thaler, R., and Tversky, A. (1997), 'The effect of myopia and loss aversion on risk taking: an experimental test', *Quarterly Journal of Economics*, 112: 647–61.

— and Tversky, A. (1979), 'Prospect theory: an analysis of decision under risk', *Econometrica*, 47: 263–91.

Kalwij, A. S. and Alessie, R. (2003), 'Permanent and transitory wage inequality of British men, 1975–2001: year, age and cohort effects', available online at: http://ssrn.com/abstract=382780.

Khandani, A. E., Lo, A. W., and Merton, R. C. (2009), 'Systematic risk and the refinancing ratchet effect', Working Paper No. 15362, Cambridge, MA: National Bureau of Economic Research.

Konzelmann, S. (2005), 'Varieties of Capitalism: Production and Market Relations in the USA and Japan', *British Journal of Industrial Relations*, 43: 593–603.

—, Wilkinson, F., Fovargue-Davies, M., and Sankey, D. (2010), 'Governance, Regulation and Financial Market Instability: The Implications for Policy', *Cambridge Journal of Economics*, 34(5): 929–54.

Krueger, J. I. and Funder, D. C. (2004), 'Towards a balanced social psychology: causes, consequences, and cures for the problem-seeking approach to social behavior and cognition', *Behavioral and Brain Sciences*, 27: 313–28.

Kuper, A. (2000), *Culture: The Anthropologists' Account*, Cambridge, MA: Harvard University Press.

Laibson, D. (2003), 'Golden eggs and hyperbolic discounting', *Quarterly Journal of Economics*, 62: 443–77.

Langley, P. (2006), 'The making of investor subjects in Anglo-American pensions', *Environment and Planning D-Society and Space*, 24(6): 919–34.

— (2008), *The Everyday Life of Global Finance: Saving and Borrowing in Anglo-America*, Oxford: Oxford University Press.

Bibliography

Langley, P. (2009), 'Consumer credit, self-discipline, and risk management', in Clark, G. L., Dixon, A. D., and Monk, A. H. B. (eds.), *Managing Financial Risks: From Global to Local* (pp. 280–300), Oxford: Oxford University Press.

Leamer, E. (2007), 'Housing is the business cycle', Working Paper No. 13428, Cambridge, MA: National Bureau of Economic Research.

Lee, R., Clark, G. L., Pollard, J., and Leyshon, A. (2009), 'The remit of financial geography—before and after the financial crisis', *Journal of Economic Geography*, 9: 723–47.

Legros, F. (2006), 'Life-cycle options and preferences', in Clark, G. L., Munnell, A. H., and Orszag, J. M. (eds.), *The Oxford Handbook of Pensions and Retirement Income* (pp. 183–200), Oxford: Oxford University Press.

Lewontin, R. C. (1993), *Biology as Ideology: The Doctrine of DNA*, New York: Harper.

Leyshon, A., Burton, D., Knights, D., Alferoff, C., and Signoretta, P. (2004), 'Towards an ecology of retail financial services: understanding the persistence of door-to-door credit and insurance providers', *Environment and Planning A*, 36: 625–45.

—————— (2006), 'Walking with moneylenders: the ecology of the UK home-collected credit industry', *Urban Studies*, 43: 161–86.

—— and Thrift, N. (1997), *Money/Space: Geographies of Monetary Transformation*, London: Routledge.

—————— and J. Pratt (1998), 'Reading financial services: texts, consumers, and financial literacy', *Environment and Planning D: Society and Space*, 16: 29–55.

Libet, B. (2004), *Mind Time*, Cambridge, MA: Harvard University Press.

Lin, Q. C. and Lee, J. (2004), 'Consumer information search when making investment decisions', *Financial Services Review*, 13: 319–32.

Litterman, B. and the Quantitative Resources Group, Goldman Sachs Asset Management (2003), *Modern Investment Management*, New York: Wiley.

Lively, J. (1979), 'Paternalism', in Griffiths, A. P. (ed.), *Of Liberty* (pp. 147–65), Royal Institute of Philosophy Lecture Series: 15, Cambridge: Cambridge University Press.

Lowenstein, R. (2004), *Origins of the Crash: The Great Bubble and Its Undoing*, New York: Penguin Press.

Lucas, R. E. (1986), 'Adaptive behavior and economic theory', *Journal of Business*, 59: S401–26.

Lusardi, A. and Mitchell, O. S. (2005), 'Financial literacy and planning: implications for retirement wellbeing', Working Paper 46/05, Center for Research on Pensions and Welfare Policies, available online at: http://cerp.unito.it.

———— (2007), 'Baby boomer retirement security: the roles of planning, financial literacy, and housing wealth', *Journal of Monetary Economics*, 54(1): 205–24.

———— (2008), 'Planning and financial literacy: how do women fare?', *American Economic Review*, 98(2): 413–17.

MacKenzie, G. A. (2006), *Annuity Markets and Pension Reform*, Cambridge: Cambridge University Press.

Madrian, B. C. and Shea, D. F. (2000), 'The Power of Suggestion: Inertia in 401(k) Participation and Savings Behavior', NBER Working Paper No. 7682, Cambridge, MA: National Bureau of Economic Research.

March, J. G. (1994), *A Primer on Decision Making*, New York: Free Press.

Markowitz, H. M. (1952), 'Portfolio selection', *Journal of Finance*, 7: 77–91.

Bibliography

May, J., Wills, J., Datta, K., Evans, Y., Herbert, J., and McIlwaine, C. (2007), 'Keeping London working: global cities, the British state and London's new migrant division of labour', *Transactions of the Institute of British Geographers*, NS32(2): 151–67.

McDowell, L. (2004), 'Work, workfare, work/life balance and an ethic of care', *Progress in Human Geography*, 28: 145–63.

—— (2005), 'Love and money: critical reflections on welfare-to-work policies in the UK', *Journal of Economic Geography*, 5(3): 365–79.

—— (2008), 'Bodies out of place: unruly female bodies in professional occupations and trans-national body shopping', Mimeo, Oxford: Oxford University Centre for the Environment.

—— (2009), 'New masculinities and femininities: gender divisions in the new economy', in Furlong, A. (ed.), *Handbook of Youth and Young Adulthood: New Perspectives and Agendas* (pp. 58–65), London: Routledge.

——, Batnitzky, A., and Dyer, S. (2008), 'Internationalization and the spaces of temporary labour: the global assembly of a local workforce', *British Journal of Industrial Relations*, 46: 750–70.

—— —— —— (2009), 'Precarious work and economic migration: emerging immigrant divisions of labour in Greater London's service sector', *International Journal of Urban and Regional Research*, 33(1): 3–25.

Mele, R. R. (2009), *Effective Intentions: The Power of Conscious Will*, Oxford: Oxford University Press.

Mellers, B. and McGraw, A. P. (2004), 'Self-serving beliefs and the pleasure of outcomes', in Brocas, I. and Cartillo, J. D. (eds.), *The Psychology of Economic Decisions. Volume 2: Reasons and Choices* (pp. 31–46), Cambridge: Cambridge University Press.

Merton, R. C. (1969), 'Lifetime portfolio selection under uncertainty: the continuous-time case', *Review of Economics and Statistics*, 51: 247–57.

—— and Bodie, Z. (2005), 'The Design of Financial Systems: Towards a Synthesis of Function and Structure', *Journal of Investment Management*, 3:1–23.

Miller, D. (2002), 'Turning Callon the right way up', *Economy and Society*, 31(2): 218–33.

Milligan, K. (2004), 'Life-cycle asset accumulation and allocation in Canada', Working Paper No. 10860, Cambridge, MA: National Bureau of Economic Research.

Minns, R. (2001), *The Cold War in Welfare: Stock Markets Versus Pensions*, London: Verso.

Mishkin, F. S. (2007), 'Housing and the monetary transmission mechanism', Discussion Paper 40, Washington, DC: Board of Governors of the Federal Reserve System.

Mitchell, O. S. and Utkus, S. P. (2004a), 'Lessons from Behavioral Finance for Pension Plan Design', in Mitchell, O. S. and Utkus, S. P. (eds.), *Pension Design and Structure: New Lessons from Behavioural Finance* (pp. 3–42), Oxford: Oxford University Press.

—— —— (eds.) (2004b), *Pension Design and Structure: New Lessons from Behavioral Finance*, Oxford: Oxford University Press.

Monk, A. H. B. (2008), 'The Knot of Contracts: The Corporate Geography of Legacy Costs', *Economic Geography*, 84: 211–36.

Mullainathan, S. (2007), 'Psychology and economic development', in Diamond, P. and Vartiainen, H. (eds.), *Behavioral Economics and its Applications* (pp. 85–113), Princeton: Princeton University Press.

Bibliography

Munnell, A. H. (2006), 'Employer-Sponsored Plans: The Shift from Defined Benefit to Defined Contribution', in Clark, G. L., Munnell, A. H. and Orszag, J. M. (eds.), *The Oxford Handbook of Pensions and Retirement Income* (pp. 359–80), Oxford: Oxford University Press.

——, Soto, M., Libby, J., and Princivalli, J. (2006), 'Investment returns: defined benefit vs. 401(k) plans', *Issue in Brief*, Boston, MA: Boston College, Centre for Retirement Research.

—— and Sundén, A. (2004), *Coming up short: the challenge of 401(k) plans*, Washington, D.C: Brookings Institution Press.

Munro, M. and Smith, S. J. (2008), 'Calculated affection? Charting the complex economy of home purchase', *Housing Studies*, 23: 349–67.

Murthi, M., Orszag, J. M., and Orszag, P. (1999), 'The value for money of annuities in the UK: theory, evidence and policy', Working Paper, London: Department of Economics, Birkbeck College.

National Audit Office (2009), *The Nationalisation of Northern Rock*, London: HM Treasury.

Nesbitt, S. and Neary, D. (2001), *Ethnic Minorities and their Pension Decisions: a Study of Pakistani, Bangladeshi and White Men in Oldham*, York: Joseph Rowntree Foundation.

Nickell, S. (2005), 'Practical issues in United Kingdom monetary policy, 2000–2005', *Proceedings of the British Academy*, 139:1–33.

Nozick, R. (1997), *Socratic Puzzles*, Cambridge, MA: Harvard University Press.

Nussbaum, M. (1994), *The Therapy of Desire*, Princeton: Princeton University Press.

Oaksford, M. and Chater, N. (2007), *Bayesian Rationality: The Probabilistic Approach to Human Reasoning*, Oxford: Oxford University Press.

O'Donoghue, T. and Rabin, M. (1999), 'Doing it now or later', *American Economic Review*, 89: 103–24.

ONS (2006*a*), 'Regional accounts highlights: household income', National Statistics, retrieved 18 July 2006, from: http://www.statistics.gov.uk/cci/nugget.asp?id=1442.

—— (2006*b*), 'Regional trends: labour market highlights', National Statistics, retrieved 17 July 2006, from: http://www.statistics.gov.uk/cci/nugget.asp?id=1435.

—— (2006*c*), 'Households with Internet access: by household type', National Statistics, retrieved 25 July 2006, from: http://www.statistics.gov.uk/StatBase/ssdataset.asp?vlnk=7203&More=Y.

Organisation for Economic Cooperation and Development (OECD) (2004), 'Global Pension Statistics Project: Measuring the Size of Private Pensions with an International Perspective', *Financial Market Trends*, No. 87, Paris.

—— (2007), 'Closing the Pensions Gap: The Role of Private Pensions', Policy Brief (September), Paris.

—— (2010), 'Assessing Default Investment Strategies in Defined Contribution Pension Plans', Paris.

Ortony, A., Clore, G. L., and Collins, A. (1988), *The Cognitive Structure of Emotions*, Cambridge: Cambridge University Press.

Pacherie, E. (2006), 'Towards a dynamic theory of intentions', in Pockett, S., Banks, W. P., and Gallagher, S. (eds.), *Does Consciousness Cause Behaviour?* (pp. 145–68), Cambridge, MA: MIT Press.

Padoa-Schioppa, C. (2008), 'The syllogism of neuro-economics', *Economics and Philosophy*, 24: 449–57.
Papke, L. (1998), 'How are participants investing their accounts in participant-directed individual account pension plans?', *American Economic Review*, 88: 212–16.
Parkinson, S., Searle, B. A., Smith, S. J., Stoakes, A., and Wood, G. (2009), 'Mortgage equity withdrawal in Australia and Britain', *European Journal of Housing Policy*, 9: 365–89.
Peck, J. (2005), 'Economic sociologies in space', *Economic Geography*, 81(2): 129–75.
—— (2010), *Constructions of neoliberal reason*, Oxford: Oxford University Press.
—— and Theodore, N. (2007), 'Variegated capitalism', *Progress in Human Geography*, 31: 731–72.
Peet, R. (2000), 'Culture, imaginary, and rationality in regional economic development', *Environment and Planning A*, 32(7): 1215–34.
Pemberton, H., Thane, P., and Whiteside, N. (eds.) (2006), *Britain's Pensions Crisis: History and Policy*, Oxford: Oxford University Press.
Pensions Commission (2004a), *Interim Report*, London: The Stationery Office.
—— (2004b), *Pensions: Challenges and Choices, The First Report of the Pensions Commission*, London: The Stationery Office.
—— (2005), *A New Pension Settlement for the Twenty-First Century: The Second Report of the Pensions Commission*, London: The Stationery Office.
Pensions Policy Institute (2009), *Retirement Income and Assets: How can Housing Support Retirement?* London.
Pettit, P. (2007), 'Neuroscience and agent-control', in Ross, D., Spurrett, D., Kincaid, H., and Stephens, L. (eds.), *Distributed Cognition and the Will* (pp. 77–91), Cambridge, MA: MIT Press.
—— and McDowell, J. (eds.) (1986), *Subject, Thought and Context*, Oxford: Oxford University Press.
Pettus, L. and Kesmodel, R.H. (2010), 'Impact of the Pension Protection Act on Financial Advice', in Clark, R.L. and Mitchell, O.S. (eds.), *Reorienting Retirement Risk Planning* (pp. 86–104), Oxford: Oxford University Press.
Polanyi, K. (1944), *The Great Transformation*, New York: Farrar & Rinehart.
Poterba, J. M. (2006), 'Annuity markets', in Clark, G. L., Munnell, A. H., and Orszag, J. M. (eds.), *The Oxford Handbook of Pensions and Retirement Income* (pp. 562–83), Oxford: Oxford University Press.
——, Rauh, J., Venti, S. F., and Wise, D. (2006), 'Lifecycle asset allocation strategies and the distribution of 401(k) retirement wealth', Working Paper No. 11974, Cambridge, MA: National Bureau of Economic Research.
Preda, A. (2004), 'The investor as a cultural figure of global capitalism', in Knorr Cetina, K. and Preda, A. (eds.), *The Sociology of Financial Markets* (pp. 141–62), Oxford: Oxford University Press.
PricewaterhouseCoopers (2006), 'Modelling UK house price uncertainty', *Economic Outlook* (November), pp. 16–23. London.
Rachlin, H. (2000), *The Science of Self-Control*, Cambridge, MA: Harvard University Press.
Reinhart, C. M. and Rogoff, K. S. (2009), *This Time is Different: Eight Centuries of Financial Folly*, Princeton: Princeton University Press.

Bibliography

Roberts, J. (2004), *The Modern Firm: Organisational Design for Performance and Growth*, Oxford: Oxford University Press.

Roy, A. D. (1952), 'Safety first and the holding of assets', *Econometrica*, 20(3): 431–49.

Ryle, G. (1949) (2000), *The Concept of Mind*, with an Introduction by D. Dennett, London: Penguin Books.

Said, E. (1978), *Orientalism*, New York: Random House.

Samuelson, P. A. (1969), 'Lifetime portfolio selection by dynamic stochastic programming', *Review of Economics and Statistics*, 51: 239–46.

Samuelson, W. and Zeckhauser, R. (1988), 'Status quo bias in decision making', *Journal of Risk and Uncertainty*, 1: 7–59.

Sass, S. (2006), 'The Development of Employer Retirement Income Plans: from the Nineteenth Century to 1980', in Clark, G.L., Munnell, A.H., and Orszag, J.M. (eds.), *The Oxford Handbook of Pensions and Retirement Income* (pp. 76–97), Oxford: Oxford University Press.

Savulescu, J. and Hope, T. (2006), 'The Elderly and Ethical Financial Decision-making', in Clark, G. L., Munnell, A. H., and Orszag, J. M. (eds.), *The Oxford Handbook of Pensions and Retirement Income* (pp. 638–57), Oxford: Oxford University Press.

Schick, F. (1997), *Making Choices: A Recasting Decision Theory*, Cambridge: Cambridge University Press.

Scholz, J. K., Seshadri, A., and Thitatrakun, S. (2006), 'Are Americans saving "optimally" for retirement?', *Journal of Political Economy*, 114(4): 607–43.

Schwanen, T. and Ettema, D. (2009), 'Coping with unreliable transportation when collecting children: examining parents' behavior with cumulative prospect theory', *Transportation Research Part A: Policy and Practice*, 43(5): 511–25.

Schwartz, B. (2005), *The Paradox of Choice: Why More is Less*, New York: Harper Perennial.

Scott, A. J. (2006), 'A perspective of economic geography', in Bagchi-sen, S. and Lawton Smith, H. (eds.), *Economic Geography: Past, Present and Future* (pp. 56–80), London: Routledge.

—— (2007), 'Capitalism and urbanization in a new key? The cognitive-cultural dimension', *Social Forces*, 85: 1465–82.

—— (2008), *Social Economy of the Metropolis: Cognitive-Cultural Capitalism and the Global Resurgence of Cities*, Oxford: Oxford University Press.

—— and Power, D. (eds.) (2004), *Cultural Industries and the Production of Culture*, London: Routledge.

Sedlmeier, P. and Gigerenzer, G. (2001), 'Teaching Bayesian reasoning in less than two hours', *Journal of Experimental Psychology: General*, 130: 380–400.

Sharpe, W. F. (2007), *Investors and Markets: Portfolio Choices, Asset Prices, and Investment Advice*, Princeton: Princeton University Press.

Shavell, S. (2007), 'Contractual holdup and legal intervention', *Journal of Legal Studies*, 36: 325–54.

Shaw, M., Thomas, B., Davey Smith, G., and Dorling, D. (2008), *The Grim Reaper's Road Map: An Atlas of Mortality in Britain*, Cambridge: Polity Press.

Bibliography

Sheshinski, E. (2008), *The Economic Theory of Annuities*, Princeton: Princeton University Press.

Shiffrin, S. V. (2000), 'Paternalism, unconscionability doctrine, and accommodation', *Philosophy and Public Affairs*, 29: 205-50.

Shiller, R. (1993), *Macro Markets: Creating Institutions for Managing Society's Largest Economic Risks*, Oxford: Oxford University Press.

—— (2000), *Irrational Exuberance*, Princeton: Princeton University Press.

—— (2002), 'Bubbles, human judgement, and expert opinion', *Financial Analysts Journal*, 58(3): 18-26.

—— (2007), 'Understanding recent trends in house prices and home ownership', Working Paper No. 13553, Cambridge, MA: National Bureau of Economic Research.

—— (2008), *The Subprime Solution: How Today's Global Financial Crisis Happened and What to Do About It*, Princeton: Princeton University Press.

Shleifer, A. (2000), *Market Inefficiency*, Oxford: Oxford University Press.

Simmons, A. J. (1979), *Moral Principles and Political Obligations*, Princeton: Princeton University Press.

Simon, H. A. (1956), 'Rational choice and the structure of the environment', *Psychological Review*, 63(2): 129-38.

—— (1978), 'Rationality as process and product of thought', *American Economic Review*, 68: 1-16.

—— (1982), *Models of Bounded Rationality*, Cambridge, MA: MIT Press.

—— (1983), *Reason in Human Affairs*, Stanford, CA: Stanford University Press.

Smith, S. J. (2005), 'States, markets and an ethic of care', *Political Geography*, 24: 1-20.

—— (2008), 'Owner-occupation: at home with a hybrid of money and materials', *Environment and Planning A*, 40(3): 520-35.

—— (2009), 'Managing financial risks: the strange case of housing', in Clark, G. L., Dixon, A. D., and Monk, A. H. B. (eds.), *Managing Financial Risks: From Global to Local* (pp. 233-57), Oxford: Oxford University Press.

—— and Easterlow, D. (2005), 'The strange geography of health inequalities', *Transactions of the Institute of British Geographers*, NS30(2): 173-90.

——, Munro, M., and Christie, H. (2006), 'Performing (housing) markets', *Urban Studies*, 43: 81-98.

——, Searle, S. A., and Cook, N. (2009), 'Rethinking the risks of owner occupation', *Journal of Social Policy*, 38: 83-102.

Standing, G. (2009), *Work after Globalization*, Cheltenham, UK and Northampton, MA: Edward Elgar.

Stango, V. and Zinman, J. (2009), 'Exponential growth bias and household finance', *Journal of Finance*, 64: 2807-45.

Stanovich, K. E. and West, R. E. (2000), 'Individual differences in reasoning: implications for the rationality debate', *Behavioral and Brain Sciences*, 23: 645-65.

Stein, J. (2009), 'Presidential address: sophisticated investors and market efficiency', *Journal of Finance*, 64: 1517-48.

Stone, K. (2005), *From Widgets to Digits: Employment Regulation for the Changing Workplace*, Cambridge: Cambridge University Press.

Bibliography

Storper, M. (2009), 'Roepke Lecture in Economic Geography Regional Context and Global Trade', *Economic Geography*, 85(1): 1–21.

Strauss, K. (2008*a*), 'Trends in occupational pension coverage in Ontario', Working Papers in Employment, Work and Finance WPG 08–02, University of Oxford.

—— (2008*b*), 'Re-engaging with rationality in economic geography: behavioural approaches and the importance of context in decision-making', *Journal of Economic Geography*, 8(2): 137–56.

—— (2008*c*), *Risk, Responsibility and Choice in UK Occupational Pensions*, DPhil Thesis, University of Oxford.

—— (2009*a*), 'Banking on property for retirement? Attitudes to housing wealth and pensions', Working Papers in Employment, Work and Finance WPG 08–04, University of Oxford.

—— (2009*b*), 'Cognition, context, and multimethod approaches to economic decision making', *Environment and Planning A*, 41(2): 302–17.

—— (2009*c*), 'Gender, risk, and occupational pensions', in Clark, G. L., Dixon, A. D., and Monk, A. H. B. (eds.), *Managing Financial Risks: From Global to Local* (pp. 258–79), Oxford: Oxford University Press.

—— and Clark, G. L. (2010), 'Geographies of UK pensions', in Coe, N. M. and Jones, A. (eds.), *Reading the Economy: the UK in the 21st Century*, London: Sage (in press).

Sundén, A. E. and Surette, B. (1998), 'Gender differences in the allocation of assets in retirement savings plans', *American Economic Review*, 88: 207–11.

Sunley, P. (2000), 'Pension exclusion in grey capitalism: mapping the pensions gap in Britain', *Transactions of the Institute of British Geographers*, NS25(4): 483–501.

Suppes, P. (2003), 'Rationality, habits, and freedom', in Dimitri, N., Basili, M., and Gilboa, I. (eds.), *Cognitive Processes and Economic Behaviour* (pp. 137–67), London: Routledge.

Taylor, J. B. (2009), *Getting Off Track: How Government Actions and Interventions Caused, Prolonged, and Worsened the Financial Crisis*, Stanford, CA: Hoover Institution.

Teece, D. (2000), *Managing Intellectual Capital*, Oxford: Oxford University Press.

Terry, R. and Gibson, R. (2010), 'Can equity release help older home-owners improve their quality of life?' in *Solutions - Lessons for Policy and Practice*, York: Joseph Rowntree Foundation.

Thaler, R. and Sunstein, C. (2003), 'Libertarian paternalism', *American Economic Review*, 93(2): 175–9.

—— —— (2008), *Nudge: Improving Decisions About Health, Wealth and Happiness*, New Haven, CT: Yale University Press.

Thane, P. (2006), 'The History of Retirement', in Clark, G. L., Munnell, A. H., and Orzsag, J. M. (eds.), *The Oxford Handbook of Pensions and Retirement Income* (pp. 33–51), Oxford: Oxford University Press.

Thomas, B., Dorling, D., and Davey Smith, G. (2010), 'Inequalities in premature mortality in Britain: observational study from 1921 to 2007', *British Medical Journal* 341: c3639.

Thoresen Review (2008), *Generic Financial Advice*, London: The Stationery Office.

Thrift, N. (2005), *Knowing Capitalism*, London: Sage.

Todd, P. M. and Gigerenzer, G. (2000), Précis of *Simple Heuristics that Make Us Smart*, *Behavioral and Brain Sciences*, 23: 665–741.

Tom, S. M., Fox, C. R., Trepel, C., and Poldrack, R. A. (2007), 'The neural basis of loss aversion in decision-making under risk', *Science*, 315: 515–18.

Trepel, C., Fox, C.R., and Poldrack, R.A. (2005), 'Prospect theory on the brain? Toward a cognitive neuroscience of decision under risk', *Cognitive Brain Research*, 23: 34–50.

Tversky, A. and Kahneman, D. (1974), 'Judgement under uncertainty: heuristics and biases', *Science*, 185: 1124–31.

—— —— (1991), 'Loss aversion in riskless choice: a reference dependent model', *Quarterly Journal of Economics*, 106: 1039–61.

Venti, S. F. and Wise, D. A. (1990), 'But They Don't Want to Reduce Housing Equity', NBER Working Paper No. 2859, Cambridge, MA: National Bureau of Economic Research.

Viceira, L. (2008), 'Life-cycle funds', in Lusardi, A. (ed.), *Overcoming the Saving Slump* (pp. 140–77), Chicago: University of Chicago Press.

Vives, X. (2008), *Information and Learning in Markets: The Impact of Market Microstructure*, Princeton: Princeton University Press.

Wagner, R. K. (2002), 'Smart people doing dumb things: the case of managerial incompetence', in Sternberg, R. J. (ed.), *Why Smart People Can Be So Stupid* (pp. 42–63), New Haven, CT: Yale University Press.

Waine, B. (2009), 'New Labour and Pensions Reform: Security in Retirement?', *Social Policy and Administration*, 43(7): 754–71.

Watson, J. and McNaughton, M. (2007), 'Gender differences in risk aversion and expected retirement benefits', *Financial Analysts Journal*, 63(4): 52–62.

Weber, E. W., Balis, A-R., and Betz, N. E. (2002), 'A domain-specific risk-attitude scale: measuring risk perceptions and risk behaviours', *Journal of Behavioral Decision Making*, 15: 263–90.

Wenger, D. (2002), *The Illusion of Conscious Will*, Cambridge, MA: MIT Press.

Wertheimer, A. (1987), *Coercion*, Princeton: Princeton University Press.

Whiteside, N. (2003), 'Historical perspectives and the politics of pension reform', in Clark, G. L. and Whiteside, N. (eds.), *Pension Security in the 21st Century: Redrawing the Public-Private Debate* (pp. 21–43), Oxford: Oxford University Press.

Williams, B. (1995), *Making Sense of Humanity and Other Philosophical Papers*, Cambridge: Cambridge University Press.

Williamson, O. (1995), *The Mechanisms of Governance*, New York: Oxford University Press.

Winter, S. G. (1986), 'Comments on Arrow and on Lucas', *Journal of Business*, 59: S427–34.

Wiseman, D. B. and Levin, I. P. (1996), 'Comparing risky decision making under conditions of real and hypothetical consequences', *Organizational Behaviour and Human Decision Processes*, 66: 241–50.

Wójcik, D. (2009), 'The role of proximity in secondary equity markets', in Clark, G. L., Dixon, A. D., and Monk, A. H. B. (eds.), *Managing Financial Risks: From Global to Local* (pp. 140–60), Oxford: Oxford University Press.

Bibliography

World Bank (1994), *Averting the Old Age Crisis: Policies to Protect the Old and Promote Growth*, New York: Oxford University Press.

Worthington, A. C. (2006), 'Predicting financial literacy in Australia', *Financial Services Review*, 15: 59–79.

Wrigley, N. (2002), *Categorical Data Analysis for Geographers and Environmental Scientists*, Caldwell, NJ: Blackburn Press.

YouGov (2006*a*), 'YouGov methodology', retrieved 25 July 2006, from: http://www.yougov.com/corporate/aboutYGMethodology.asp?jID=1&sID=1&UID.

—— (2006*b*), 'Questions and answers', retrieved 18 July 2006, from: http://www.yougov.com/corporate/aboutQA.asp?jID=1&sID=1&UID.

Author Index

Includes all referenced authors.

Abolafia, M. 34
Agarwal, S. 71
Ainslie, G. 23, 26, 60, 87, 123, 133
Akerlof, G. A. 15
Aldrich, J. H. 48n12
Alessie, R. 83, 84
Allen, F. 21
Ambachtsheer, K. 166
Amemiya, T. 48n12
Ameriks, J. 18
Antolin, P. 162
Arber, S. 42, 57
Arrow, K. 21, 79
Astuti, R. 37
Atiyah, P. S. 141
Atkinson, A. 74, 169
Audi, R. 19
Axelrod, R. 20

Bajtelsmit, V. L. 42, 49, 70, 128
Banks, J. 104n4
Bao, Y. Q. 130
Barnes, T. J. 4
Baron, J. 20, 26n9, 29, 45, 71, 79, 80n3, 107, 122
Barro, R. 119
Bathelt, H. 17, 60
Baumeister, R. F. 30
Beck, U. 2
Beck-Gernsheim, E. 2
Bellamy, K. 57
Benartzi, S. 54, 65, 73, 89, 100, 136, 138, 149
Bermúdez, J. L. 19, 22
Bernasek, A. 42
Bernheim, D. 41, 74, 169
Bertrand, M. 42, 123, 130
Bikker, J. 159
Blair, Tony 154
Bodie, Z. 150, 156
Borio, C. 119
Bourdieu, Pierre 12
Boyer, R. 59, 82
Bratman, M. E. 61, 63, 64
Bridgen, P. 169

Brown, S. K. 161
Byrne, A. 68n5

Cain, D. M. 158n4
Callon, M. 38n15
Camerer, C. 25
Cannon, E. 119n1, 125, 126n7
Carroll, C. D. 49
Caudill, S. 48n12
Charles, N 2
Chater, N. 27, 107
Choi, A. 158n4
Choi, J. J. 41
Christie, H. 77
Church, B. K. 158n4
Clark, G. L. 11, 13, 21n4, 21n5, 22n6, 28n12, 29, 30, 33, 40, 41, 43n6, 45, 49, 50, 53, 61, 65, 67, 79, 80n3, 81, 82n6, 83, 85, 90, 91, 99, 100, 104, 115, 120, 121, 122, 124, 127, 128, 133, 136, 137, 139, 141, 143, 147, 148, 150, 155, 157, 159, 160n7
Clark, R. L. 164
Compton, J. 56
Conefrey, T. 97
Conway, N. 155
Cooper Review 168
Costa, D. L. 56
Couclelis, H. 25, 27, 31
Coval, J. D. 104
Cremers, M. 157
Cronqvist, H. 58n16
Cutler, N. E. 57, 169
Cyert, R. 21

Damasio, A. R. 61n2
Damasio, H. 61n2
Deaton, A. 59, 71, 87
De Deken, J. J. 59, 99, 120
Department for Work and Pensions (DWP) 39, 74, 128, 131–2, 154
Department of Social Security 2
Dickens, R. 83, 84, 85, 87
Disney, R. 104n4
Dixon, A. D. 155

Author Index

Doherty, M. E. 27, 30, 59n1, 61, 73, 122
Dohmen, T. 65, 145
Dolvin, S. D. 41
Dorn, D. 42n5, 145
Døskeland, T. M. 80n3
Douglas, M. 29
Dreu, J. de 159
Dworkin, G. 141

Easterlow, D. 25, 37, 59, 121
Ebbinghaus, B. 33
Engelen, E. 26, 82
Esping-Andersen, G. 59
Ettema, D. 25

Flaherty, C. 164
Foucault, M. 38n15
Frericks, P. 2
Friedberg, L. 30, 42, 56, 72
Fries, J. F. 130n12
Funder, D. C. 26n9, 30, 40, 107, 132

Gabaix, X. 62, 72, 74, 79, 139, 149, 159n5
Gage, Phineas 61
Gale, D. 21
Gardner, J. 125
Gerald, J. F. 97
Gertler, M. S. 37, 77, 79–80, 116
Ghilarducci, T. 136, 139
Gibson, R. 98
Gigerenzer, G. 21, 22, 26, 37, 41, 61, 73, 119, 132
Ginn, J. 1, 42, 57
Glennie, P. 25, 38, 61
Glückler, J. 17, 60
Goda, G. S. 164
Goffman, E. 38n15
Goldstein, W. M. 73
Goldsticker, R. 72n8
Goos, M. 83n7
Granovetter, M. 17
Greenspan, Alan 3n1

Hägerstrand, T. 27
Hallahan, T. A. 42, 76, 122
Harré, R. 24
Harris, C. 2
Harris, P. L. 37
Heaton, J. 87
Henrich, J. 7n3, 18, 42, 123, 130
Hershey, D. 61, 123
Hills, J. 107, 117
Hilton, D. J. 81
Hogarth, R. M. 6, 23, 28, 37, 38
Hollis, M. 42n5
Holmes, S. 141
Hong, H. 34

Hope, T. 165
Huberman, G. 26n10, 42n5, 145
Hurley, S. 20, 24, 62
Hvide, H. K. 80n3

Iacobucci, E. 159n5
Inman, P. 170
Iyengar, S. 5, 41, 101

Jackson, H. 156
Jacobs-Lawson, J. M. 123
Jianakoplos, N. 128

Kahn, M. E. 56
Kahneman, D. 7, 18, 20, 21, 23, 25, 27, 32, 41, 49, 59, 61, 73, 79, 90n14, 101, 106, 110, 123n3, 124, 131, 132, 138
Kalwij, A. S. 83, 84
Kesmodel, R. H. 162
Khandani, A. E. 104n5
Knox-Hayes, J. 127, 128, 137, 147, 148
Konzelmann, S. 155
Krueger, J. I. 26n9, 30, 40, 107, 132
Kuang, X. 158n4
Kuper, A. 17

Laibson, D. 87, 133, 159n5
Langley, P. 3, 16, 26n8, 59, 81, 82, 90n13, 98, 118, 137, 151
Leamer, E. 86
Lee, J. 131
Lee, R. 3n1, 115
Legros, F. 56
Levin, I. P. 123
Lewontin, R. C. 20n3
Leyshon, A. 32n14, 42n4, 121, 129, 130n11, 132, 133, 169
Libet, B. 61, 64
Lin, Q. C. 131
Litterman, B. 90
Lively, J. 141
Lowenstein, R. 119, 124
Lucas, D. 87
Lucas, R. E. 21
Lusardi, A. 31n13, 41, 42, 60, 101, 169

McDowell, J. 38
McDowell, L. 4, 27, 28, 57, 83n7, 84n8
McGraw, A. P. 105
MacKenzie, G. A. 126n7, 130
McNaughton, M. 91
Madrian, B. C. 101
Manning, A. 83n7
March, J. 21
Markowitz, H. M. 89
May, J. 83n7, 97, 98, 105
Mele, R. R. 64n3

Author Index

Mellers, B. 105
Merton, R. C. 87, 156
Meyer, T. 169
Mill, John Stuart 3, 21n4, 150
Miller, D. 25
Milligan, K. 49
Minns, R. 118
Mitchell, O. S. 18, 31n13, 41, 42, 60, 101, 151, 169
Monk, A. 40, 82n6, 83, 120, 121, 155
Moskowitz, T. J. 104
Mullainathan, S. 81
Munnell, A. H. 1, 41, 59, 120, 136
Munro, M. 103
Murthi, M. 125n4

National Audit Office 76
Neary, D. 130
Nelson, F. D. 48n12
Nesbitt, S. 130
Nickell, S. 76
Nozick, R. 140

Oaksford, M. 27, 107
O'Connor, K. 104
O'Donoghue, T. 96
OECD 50n13, 150, 156n1
ONS 125n5, 133
Orszag, J. M. 125
Ortony, A. 8n4

Pacherie, E. 64
Padoa-Schioppam, C. 18n1
Papke, L. 49
Parkinson, S. 103n3
Peck, J. 3, 17, 25
Peet, R. 4
Pemberton, H. 128
Pensions Commission 39n1, 43n6, 64, 97n1, 102, 120, 137, 150
Pensions Policy Institute 98n2, 103n3
Pettit, P. 38, 152
Pettus, L. 162
Polanyi, K. 25
Pollack, R. A. 56
Poterba, J. M. 49, 64, 104, 119n1, 139
Power, D. 26
Preda, A. 59, 96, 140
PricewaterhouseCoopers 76

Rabin, M. 96
Rachlin, H. 23, 133n13, 138
Rake, K. 57
Reinhart, C. M. 97, 104
Roberts, J. 83, 155
Rogoff, K. S. 97, 104

Roy, A. D. 94, 106
Ryle, G. 64

Said, E. 20
Samuelson, P. A. 87
Samuelson, W. 139
Savulescu, J. 165
Schick, F. 22
Scholz, J. K. 41, 59
Schwanen, T. 25
Scott, A. J. 25, 26, 28
Sedlmeier, P. 26
Sharpe, W. F. 89, 90, 96, 102
Shavell, S. 159n5
Shea, D. F. 101
Sheshinski, E. 65, 72n8
Shiffrin, S. V. 141
Shiller, R. 15, 27, 76, 79, 81, 88n11, 114, 115, 118, 139
Shleifer, A. 79
Shwiff, S. 42
Simmons, A. J. 143, 144
Simon, Herbert 8, 18, 21, 25, 41, 79, 96, 101, 131, 138
Smith, S. J. 25, 32, 37, 59, 82, 89n12, 95, 98, 103, 105, 121
Standing, G. 1, 2
Stango, V. 104n5
Stanovich, K. E. 131, 142
Stein, J. 10n5, 78
Storper, M. 38
Strauss, K. 4, 15, 23, 38n15, 40, 41, 50, 53, 59, 60, 65, 82, 95, 96, 98, 101, 117, 121, 122, 137, 154, 170
Sundén, A. 1, 50, 136
Sunley, P. 67, 119, 130n11
Sunstein, C. 28, 33, 58, 101, 137, 151, 162, 164
Suppes, P. 24
Surette, B. 50

Taylor, J. B. 86
Teece, D. 83
Templeton, W. K. 41
Terry, R. 98
Thaler, R. 28, 33, 54, 58, 65, 73, 89, 100, 101, 136, 137, 138, 149, 151, 162, 164
Thane, P. 1
Theodore, N. 25
Thoresen Review 73
Thrift, N. 25, 38, 61
Todd, P. M. 37
Tom, S. M. 111n8
Tonks, I. 119n1, 125, 126n7
Trepel, C. 107
Treussard, J. 150
Triantis, G. 158n4

Author Index

Tversky, A. 7, 18, 21, 23, 25, 32, 41, 59, 74, 79, 90n14, 101, 106, 110, 124, 131, 138

Urwin, R. 157, 159
Utkus, S. P. 41, 151

Venti, S. F. 104n4
Viceira, L. 162
Vives, X. 27

Wagner, R. K. 21n5, 61, 74
Watson, J. 91
Webb, A. 30, 42, 56, 72
Weber, E. W. 41, 42n5, 145
Wenger, D. 62, 72n6
Wertheimer, A. 142
West, R. E. 131, 142

Whitelegg, C. 76n1
Whiteside, N. 21n4, 33, 121, 150
Wildavsky, A. 29
Williams, Bernard 4, 10n6, 29
Williamson, O. 155
Winter, S. G. 21, 22n6
Wise, D. A. 104n4
Wiseman, D. B. 123
Wójcik, D. 26n10
World Bank 99
Worthington, A. C. 131
Wrigley, N. 110

YouGov 125

Zeckhauser, R. 139
Zinman, J. 104n5

Subject Index

active choice, and pension delivery
   design 162-3
adequacy:
   and pension systems 154
   and public utility model 170
advertising, and financial decision-making 13
advisory services, and planning 74
affordability, and pension reform 154
age:
   and annuities 129
   and asset allocation 100-1
   and home ownership as form of retirement
      saving 109-10, 113
   and number of supplementary
      pensions 127, 131
   and planning 67, 68, 95
   and risk propensities 49, 51, 56
ageing populations 2
agency, and pension delivery design 169
annuities 32
   and decline in rates 124-5
   and definition of 119n1
   and demand for 128-30
   and impact of economic downturn 124-5
   and knowledge of 60, 70
   and region of residence 119, 128-9, 130,
      131, 132
   and resistance to 104
   and socio-demographic characteristics 70,
      128-9, 131
   and switching behavior 129-30
   and willingness to take 126
anthropology, and behavior 17
asset allocation:
   and consultation 137, 144-5
      expressed preference for 146-7, 148-9
      no expressed preference for 148
   and defined contribution pensions 100-1
   and financial competence 139
   and gender 53
   and socio-demographic characteristics 52-5
   and target-date funds 139-40
attention 139
Australia, and pension system 123, 158,
   166-9
authoritarianism, and new paternalism 152

auto-enrolment 137, 140
   and coercion 150
   as default setting 161
   and employee pensions 64, 65
   and new paternalism 142, 149-50

Basic State Pension (BSP) 154
   and declining value of 39
   and significance of 39
behavior:
   and approach to 4
   and behavioralism 20-1
   and context 18, 28, 96
   and deliberation 27
   and dualistic explanations of 17
   and environment 17-18, 28
   and geographical scale of 24-5
   and habit 23-4
   and Herbert's scissors metaphor 8, 18
   and human nature 19-22
   and imitation 24
   and institutional context of experiments
      on 29
   and intuition 23
   and local focus of 26-7
   and myopia 26
   and reflection 27-8
   and uneven distribution of cognitive
      skills 169
behavioral finance, and critique of rational
   actor model 3
behavioral revolution 41-2
   and myopia 79
   and new paternalism 151
behavioralism, and decision-making 20-1
beliefs, and planning 63-4
bequest motive, and resistance to
   annuities 103-4
Beveridge Report (1944) 150
biology, and behavior 20
Blackstone Global Investors (BGI) LifePath
   Portfolio 139-40
bundling 159n5

Chicago school 33
children, and financial education 16

Subject Index

choice:
    and habit 24
    and supplementary pensions 171
coercion 33
    and auto-enrolment 150
    and definition of 142
    and new paternalism 142
    and paternalism 142
cognition, and meaning of 37
cognitive science, and critique of rational actor model 3
cognitive skills, and uneven distribution 169
commitment, and employers 155–6
communication, and pension delivery design 165–6
compensation packages:
    and contingent compensation 83–4
    and finance sector 83
confidence 23
    and demand for consultation 147
consultation:
    and asset allocation 137, 144–5
    and correlates of expressed preference for 146–7, 148–9
    and correlates of no expressed preference for 148
    and income 145–6
    and new paternalism 145, 149, 152
    and survey questions on 144–5
context:
    and behavior 18, 28, 96
    and decision-making 25–6, 122
    and future research on 169–70
    and meaning of 37–8
    and rationality 7
    and tacit knowledge 77
contingent behavior 123–4
contingent compensation 83–4
contract of employment 2
contract theory 159
contribution escalators 138, 162
Cooper Review (Australia) 167–8
cultural capital 12
culture, and behavior 17

decision theory 19
decision-making:
    and behavioralism 20–1
    and cognitive performance 138
    and consistency of 60–1
    and context 25–6, 122
    and contingent behavior 123–4
    and deliberation 27
    and domain-specific skills 61
    and habit 23–4
    and heuristics 22–3, 122, 132–3
    and imitation 24, 132
    and intuition 23, 132
    and local focus of 26–7
    and logical process of 22
    and myopia 26
    and nature of 22, 132
    and practice of 25–9
    and reflection 27–8
    and reinforcement 133
    and satisficing 22, 27
    and shortcomings 138
    and social relationships 30
    and social roles 30
    see also financial decision-making
defined benefit (DB) pensions 120–1
    and alignment of employer/employee interests 154–5
    as anachronism 155
    and characteristics of 100
    and decline of 136
    as drain on company resources 155
    not valued by younger workers 155
    and post-war expansion of 100
    as preserve of the few 170
    and viability of undercut 40
defined contribution (DC) pensions:
    and asset allocation 100–1
    and characteristics of 100
    and default settings 101
    and discounted value of 101–2
    and diversity among participants 160–1
    and employees view of 155–6
    and employers view of 155, 156
    and final value of 121
    and financial decision-making 101
    and growth of 40, 120
    and individual responsibility 41
    and low savings rates 101
    and multi-employer plans 156
    and naïvity of average participant 102
    and realignment of risk from employer to employee 41
    and slow uptake of 136
    and sub-optimal decisions 41
    and United States 136–7
    see also pension delivery design
deliberation:
    and decision-making 27
    and planning 62–3
    and socio-demographic characteristics 30–1
deregulation 3
design of pension delivery, *see* pension delivery design
desires, and planning 63–4

earned income:
    and contingent compensation 83–4
    and growth rates 84–5

Subject Index

link to house price bubble 85–7
economic geography:
  and behavior 17
  and critique of rational actor model 4
economic inequality 83, 96, 170–1
  and pension systems 154
economic sociology, and behavior 17
education:
  and home ownership as form of retirement saving 110, 113, 114
  and number of supplementary pensions 127, 131
education, financial 16
  and financial decision-making 12
  and need for programs 57
  and socio-demographic characteristics 57–8
efficient markets hypothesis 78
emotional knowledge 14–15
emotions, and financial decision-making 8
engagement, and pension delivery design 164–5
environment:
  and behavior 17–18, 28
  and meaning of 37
  see also context
equity, and pension systems 154
equity release 103
European Union, and pension systems 154
evolution:
  and human nature 19–20
  and planning trait 61
experience, and financial decision-making 124

fairness, and pension reform 154
family home, as source of retirement income 98, 116–17
  and age 109–10, 113
  and arguments for importance of:
    nature of risk more easily understood 104–5
    separability from financial markets 104
  and bequest motive 103–4
  and education 110, 113, 114
  and equity release 103
  and marital status 110, 113
  and nature of family home 103
  and rental of unused rooms 103
  and risk aversion 98
  and self-insurance 103
  and socio-demographic characteristics 109–14, 116–17
  and statistical analysis and results of survey 107–13
  and survey design and risk measures 105–7
  and synthesis and interpretation of survey results 113–15

family structures, and changes in 2
finance, and geography of 81–2
finance industry, and contingent compensation 83–4
financial decision-making:
  and advertising 13
  and application of existing decision templates 6
  and competences required for 138–9
  and deep decisions 5–6
  and defined contribution pensions 101
  and education 12
  and emotional knowledge 14–15
  and emotions 8
  and environment 132
  and experience 124
  and importance of 15–16
  and information collection 6
  and instrumental value of 6
  and knowledge 6, 124
  and nature of 132
  and neoliberalism 4
  and networks 12–13
  and optimists 122
  and peer-imitation 13
  and pessimists 122
  and quality of 61
  and rational actor model 4
    critiques of 3–4, 122
  and region of residence 132
  and shallow decisions 5, 6
  and social identity 8
  and social position 124
  and social processes structuring field 171
  and social relationships 12–13
  and socio-demographic characteristics 12, 18, 122–3, 131, 132
  and subjective expected utility maximization model 7
  and uneven distribution of skills 169
  see also decision-making
financial literacy:
  and distribution of 11
  and financial education 16
  and importance of 131
  and individual responsibility 151
  and pensions 121
financial planning, *see* planning

gender:
  and annuities 129–30
  and asset allocation 53
  and demand for consultation 147
  and number of supplementary pensions 127, 128, 131
  and planning 67
  and risk propensities 49, 51, 56

197

Subject Index

geography:
   and local risk cultures 133
   and responses to global financial volatility 119
   see also region of residence
Germany 130
global financial crisis, and rational actor model 3n1
globalization, and impact of 2

habit:
   and choice 24
   and decision-making 23–4
   and geographical scale of 24–5
herd behavior 15
   and market behavior 81
heuristics, and decision-making 21, 22–3, 122, 132–3
housing, *see* family home; property
human nature:
   and behavior 19–22
   and biological model of 20
   and evolutionary model of 19–20
   and selfish model of 19
   and sociable model of 19

imitation:
   and decision-making 24, 132
   and geographical scale of 24–5
   and planning 62
income:
   and demand for consultation 145–6, 147
   and number of supplementary pensions 127–8, 131
   and planning 67, 68, 95
   and risk propensities 49, 51
individual responsibility:
   and defined contribution pensions 41
   and financial literacy 151
   and internalization of norms of 82
   and pension planning 59, 102
   and pension policy 150
   and pension reform 154
   and planning 67–8
   and presumption of 136
inertia 139
information:
   inefficient use of 11
   and Internet as source of 13–14
   and naïve investors 9–10
   and sophisticated investors 10–11
   and workplace information exchange 13
information sources:
   and doubtful utility of generic advisory services 74
   and planning 68, 69, 72–3, 74
institutional design, *see* pension delivery design

intention, and planning 63, 64
Internet, as source of financial information 13–14
intimate relationships:
   and planning 74
   and planning for retirement 34
intuition:
   and decision-making 23, 132
   and geographical scale of 24–5
   and nature of 14
   and planning 62–3

knowledge:
   and context of knowing 12–14
   and emotional knowledge 14–15
   and financial decision-making 6, 124
   and rationality 7–8

labor market, and stratification of 28
legal cases:
   Jones v Harris Associates (2008, USA) 160n7
   Renfro v Unisys Corporation et al (2010, USA) 157n3
   Tibble v Edison International et al (2010, USA) 157n3
liberalism:
   and contrast with neoliberalism 3
   and new paternalism 151
   see also neoliberalism
libertarian paternalism 33, 58, 140
   and elements of 137
   see also new paternalism; paternalism
local risk cultures 133
loss aversion 106–7, 111
love 15

marital status:
   and home ownership as form of retirement saving 110, 113
   and number of supplementary pensions 131
   and risk propensities 50, 56
market behavior:
   and behavioral predispositions 80–1
   and efficient markets hypothesis 78
   and herd behavior 81
   and myopia 79, 96
   and naïve planners 81, 96
   and opportunists 81, 96
   and rational expectations 78, 79
   and sophistication 80–1, 96
   and tacit knowledge 79
   and taxonomy of 78–81
market economies:
   and change 25
   and spatial configurations 25
Mercer Human Resource Consulting 43, 88, 144

Subject Index

methodology:
  and family home as source of retirement income 105–7
  and risk propensities:
    analysis of survey data 44–7
    survey method and implementation 43–4
  and scary market survey 125–6
  and survey on planning attitudes 65–7
migration, and tensions around 2
moral worthiness, and pension policy 150
multi-employer pension plans 167
myopia:
  and behavioral revolution 79
  and decision-making 26
  and market behavior 96

naïve investors, and financial information 9–10
naïve planners, and market behavior 81, 96
National Employment Savings Trust (NEST) 58, 120, 168, 169, 170
National Pensions Savings Scheme (NPSS) 39–40, 58, 120, 132, 137
neoliberalism:
  and contrast with liberalism 3
  and economic theory 3
  and financial decision-making 4
  and unravelling of social compact 2
  see also liberalism
networks, and financial decision-making 12–13
new paternalism 137, 138
  and assumptions of 138
  and authoritarianism 152
  and auto-enrolment 142, 149–50
  and challenge facing 152
  and coercion 142
    definition of 142
  and consultation 145, 149, 152
  and contradictions in 153
  and definition of paternalism 141
  and expectations of participants 151–2
  and impact of behavioral revolution 151
  and liberalism 151
  and obligations of planners 143, 151
  and opting-out 142, 143
  and overriding of individual decision-making 150–1
  and relationship between planners and beneficiaries 143
  and tacit consent 143–4
  see also libertarian paternalism; paternalism
Northern Rock 76
nudging, and libertarian paternalism 137

Objective Research 43–4
occupational pensions:
  and decline in scope and adequacy of 2–3
  and doubtful future importance of 120
  and employment model 2
  and increasing emphasis on importance of 2
  and multi-pillar system 154
  as pillar of retirement saving 99
  and policy commitment to 2
  and pressures on 100
  as recent innovation 1
  and workforce management and control 1
Old Age Pension Act (1908) 150
opportunists, and market behavior 81, 96
opting-out 140, 141
  and coercion 142
  and new paternalism 143
orientalism 20

paternalism:
  and coercion 142
  and definition of 141
  and nature of 140–1
  see also libertarian paternalism; new paternalism
peer-imitation, and financial decision-making 13
pension delivery design 156–7
  and active choice 162–3
  and agency problem 169
  and Australian system 166–9
  and default settings 161–2
    auto-enrolment 161
    contribution escalators 162
    default strategy 161–2
    second-level 162, 163
  and engagement 164–5
  and executive entity 158–9
  and gates 163
  and governing entity 157–8
  and hurdles 163
  and public utility model 167–9
  and public-private partnerships 166–7
  and reporting and communication 165–6
  and retirement services 165
  and service providers 159–60
pension institutions 33
Pension Protection Act (2006, USA) 136, 151, 162
Pension Protection Fund (PPF) 100, 121
pension systems:
  and adequacy 154
  and reform criteria 154
  and sustainability 154
  see also pension delivery design
people's pension 170–1
permanent income hypothesis 87
personal pension saving 120
  as pillar of retirement saving 99

Subject Index

personal pension saving (cont.)
  and requirements for effective planning 102
personhood, and reconceptualizing 34–6
planning:
  and analysis of 60
  and annuities, knowledge of 60, 70
  and beliefs and desires 63–4
  and cognitive performance 138
  and deliberation 62–3
  and difficulties with 59–60
  as human trait 61
  as illusion 62
  and imitation 62
  and individual responsibility 59, 67–8
    internalization of norms of 82
  and information sources 68, 69, 72–3, 74
  and intention 63, 64
  and intimate relationships 34
  and intuition 62–3
  and planning process 62–5
  and premium on 59
  and public policy 74
  and rationality 60–2
  and replacement rate of income supplements 67, 68–9
  and risk propensities 71
  and salience 62–3, 71–2, 74, 95
    diversity in recognition of 160–1
  and scale of 72–3, 74
  and shortcomings 138
  in social science literature 61–2
  and socio-demographic characteristics 30–1, 34, 70–1, 107–8
    age 67, 68, 95
    gender 67
    income 67, 68, 95
    region of residence 67, 68
    spouse's pension 68, 69
  and sophistication 62–3, 71, 73–4, 95
  and survey questions 65–7
  summary of findings 67–70
portfolio diversification 94
  and determinants of 93, 95
  and property 82–3, 90–1
    in combination with other instruments 92
    determinants of bias toward property 91–2, 93–4, 96
  and retirement investment 87–90
  and safety-first strategy 94, 95
  and socio-demographic characteristics 89, 93–4
portfolio management, and retirement saving 102
privatization 121
professional qualifications, and financial decision-making 12

property:
  and attractiveness as investment 97, 98, 103
  and housing boom 76, 77, 82
    evidence of housing bubble 84–7, 95, 97–8
  and investment portfolios 77–8
  and portfolio diversification 82–3, 90–1
    determinants of bias toward property 91–2, 93–4, 96
    property with other instruments 92
  and retirement saving 32, 82
  and supply shortfall 76
  see also family home
prospect theory 32, 107
public utility model:
  and adequacy 170
  and pension delivery design 167–9
  and sustainability 170–1
public-private partnerships, and pension delivery design 166–7

rational actor model 3
  and critiques of 3–4, 122
  and financial decision-making 4
  and global financial crisis 3n1
rationality:
  and conceptions of 9
  and context 7
  and knowledge 7–8
  and planning 60–2
  and subjective expected utility maximization model 7–8
reflection, and decision-making 27–8
region of residence:
  and annuities 119, 128–9, 130, 131, 132
  and demand for consultation 147
  and financial decision-making 132
  and number of supplementary pensions 126–7, 131
  and planning 67, 68
regional inequality, and local risk cultures 133
reinforcement, theory of 133n13
replacement rate of income supplements, and planning 67, 68–9
reporting, and pension delivery design 165–6
representative agent 34–6
retirement investment:
  and portfolio diversification 87–90, 94
    determinants of 93, 95
    determinants of bias toward property 91–2, 93–4, 96
    property 82–3, 90–1
    property with other instruments 92
  and safety-first strategy 94, 95
retirement saving:
  as exercise in portfolio management 102
  and family home as source of retirement income 98

200

and housing 101–5
and requirements for effective planning 102
and three pillars of 99–101, 120
retirement services, and pension delivery design 165
risk 9
  and collectivization of 171
  and individualization of 169
  and uncertainty 138–40
risk assessment, and socio-demographic characteristics 30
risk aversion 7, 11
  and age 49, 51, 56
  and gender 49, 51, 56
  and income 49, 51
  and marital status 50, 56
  and reliance upon family home for retirement income 98
  and reliance upon property for retirement income 98
  and socio-demographic characteristics 12
    marginal effects 50–2
  and spouse's pension 50, 51–2, 56–7
risk propensities:
  and cross-national studies 123
  and family home as source of retirement income 116
    statistical analysis and results of survey 107–13
    survey design and risk measures 105–7
    synthesis and interpretation of survey results 113–15
  and geography 132
  and local culture 130, 132, 133
  and planning 71
  and socio-demographic characteristics 40, 41–2, 122–3
    age 49, 51, 56
    analysis of survey data 44–7
    asset allocation 52–5
    gender 49, 51, 56
    implications of relationship between 57–8
    income 49, 51
    marginal effects on risk aversion 50–2
    marital status 50, 56
    relationship between 47–9, 55–7
    spouse's pension 50, 51–2, 56–7
    survey method and implementation 43–4
salience:
  and diversity in recognition of 160–1
  and planning 62–3, 71–2, 74, 95
satisficing 79
  and decision-making 22, 27
Save More Tomorrow 136
scale:
  and pension planning 74

  and planning 72–3
scary market survey 125–6
  and annuities:
    demand for 128–30
    willingness to take 126
  and implications and conclusions 131–3
  and number of supplementary pensions 125–6
    correlates of 126–8
    factors affecting 126
self-directed pension savings, and financial competence 139
simplicity, and pension reform 154
skills, and labor market stratification 28
social compact 2
social identity, and financial decision-making 8
social polarization 154, 170–1
social position, and financial decision-making 124
social relationships:
  and decision-making 30
  and financial decision-making 12–13
social roles, and decision-making 30
social security, as pillar of retirement saving 99
social sorting, and partner pension entitlement 56–7
socio-demographic characteristics:
  and annuities 131
    demand for 128–9
    knowledge of 70
  and asset allocation 52–5, 100–1
  and consultation:
    expressed preference for 146–7, 148–9
    no expressed preference for 148
  and deliberation 30–1
  and financial decision-making 12, 18, 122–3, 131, 132
    uneven distribution of skills 169
  and financial education 57–8
  and home ownership as form of retirement saving 109–14, 116–17
  and number of supplementary pensions 126–8, 131
  and planning 30–1, 34, 70–1, 107–8
    age 67, 68, 95
    gender 67
    income 67, 68, 95
    region of residence 67, 68
    spouse's pension 68, 69
  and portfolio diversification 89, 93–4
  and risk assessment 30
  and risk groups 52
  and risk propensities 40, 41–2, 122–3
    age 49, 51, 56
    analysis of survey data 44–7
    asset allocation 52–5

201

Subject Index

socio-demographic characteristics: (cont.)
    gender 49, 51, 56
    implications of relationship between 57–8
    income 49, 51
    marginal effects on risk aversion 50–2
    marital status 50, 56
    relationship between 47–9, 55–7
    spouse's pension 50, 51–2, 56–7
    survey method and implementation 43–4
sophisticated investors, and financial information 10–11
sophistication:
    and market behavior 80–1, 96
    and planning 62–3, 71, 73–4, 95
South Africa 130
spouse's pension:
    and demand for consultation 147
    and planning 68, 69
    and risk propensities 50, 51–2, 56–7
state pensions 120
    as pillar of retirement saving 99
    and rationale for 1
    see also Basic State Pension (BSP)
stock markets:
    and correlation with income and property prices 85–7
    and volatility of 118–19
subjective expected utility (SEU):
    and financial decision-making 7
    and rationality 7–8
supplementary pensions 120
    and choice 171
    and correlates of 126–8
    and number of 125–6
        factors affecting 126

    and region of residence 126–7, 131
    and socio-demographic characteristics 126–8, 131
sustainability:
    and pension systems 154
    and public utility model 170–1

tacit consent, and new paternalism 143–4
tacit knowledge:
    and context 77
    and market behavior 79
target-date funds 139–40
technology, media and telecommunications (TMT) bubble 118, 131
    and annuity take-up 126
Towers Watson 125, **157**
training, and financial decision-making 12
trust 34
Turner Report (Pensions Commission, 2005) 120, 131–2, 137, 150
    and recommendations of 39
    as response to pensions mis-selling scandal 121

uncertainty, and risk 138–40
United States:
    and defined contribution pensions 136–7
    governance 157–8
    and pension regulation 151
universal citizen's pension 154
utility maximization 3

workplace information exchange 13

YouGov 125